Romantic Automata

TRANSITS:
LITERATURE, THOUGHT & CULTURE, 1650–1850

A long running and landmark series in long eighteenth-century studies, *Transits* includes monographs and edited volumes that are timely, transformative in their approach, and global in their engagement with arts, literature, culture, and history. Books in the series have engaged with visual arts, environment, politics, material culture, travel, theater and performance, embodiment, writing and book history, sexuality, gender, disability, race, and colonialism from Britain and Europe to the Americas, the Far East, and the Middle East. Proposals should offer critical examination of artifacts and events, modes of being and forms of knowledge, material culture, or cultural practices. Works that make provocative connections across time, space, geography, or intellectual history, or that develop new modes of critical imagining are particularly welcome.

Recent titles in the *Transits* series:

Romantic Automata: Exhibitions, Figures, Organisms
Michael Demson and Christopher R. Clason, eds.

Beside the Bard: Scottish Lowland Poetry in the Age of Burns
George S. Christian

The Novel Stage: Narrative Form from the Restoration to Jane Austen
Marcie Frank

The Imprisoned Traveler: Joseph Forsyth and Napoleon's Italy
Keith Crook

Fire on the Water: Sailors, Slaves, and Insurrection in Early American Literature, 1789–1886
Lenora Warren

Community and Solitude: New Essays on Johnson's Circle
Anthony W. Lee, ed.

The Global Wordsworth: Romanticism Out of Place
Katherine Bergren

Cultivating Peace: The Virgilian Georgic in English, 1650–1750
Melissa Schoenberger

Jane Austen and Comedy
Erin M. Goss, ed.

Intelligent Souls?: Feminist Orientalism in Eighteenth-Century English Literature
Samara Anne Cahill

The Printed Reader: Gender, Quixotism, and Textual Bodies in Eighteenth-Century Britain
Amelia Dale

For a full list of *Transits* titles, please visit our website: www.bucknell.edu/UniversityPress.

Romantic Automata

EXHIBITIONS, FIGURES, ORGANISMS

Edited by
MICHAEL DEMSON AND
CHRISTOPHER R. CLASON

Bucknell University | UNIVERSITY PRESS
LEWISBURG, PENNSYLVANIA

Library of Congress Cataloging-in-Publication Data

Names: Demson, Michael, editor. | Clason, Christopher R., editor.
Title: Romantic automata : exhibitions, figures, organisms / edited by Michael Demson, Christopher R. Clason.
Description: Lewisburg, Pennsylvania : Bucknell University Press, 2020. | Series: Transits: literature, thought & culture 1650–1850 | Includes bibliographical references and index.
Identifiers: LCCN 2019028588 | ISBN 9781684481767 (paperback) | ISBN 9781684481774 (hardback) | ISBN 9781684481781 (epub) | ISBN 9781684481798 (mobi) | ISBN 9781684481804 (pdf)
Subjects: LCSH: English literature—19th century—History and criticism. | English literature—18th century—History and criticism. | Robots in literature. | Literature and technology—Great Britain—History—19th century. | Literature and technology—Great Britain—History—18th century. | Romanticism—Great Britain.
Classification: LCC PR457 .R626 2020 | DDC 820.9/145—dc23
LC record available at https://lccn.loc.gov/2019028588

A British Cataloging-in-Publication record for this book is available from the British Library.

∞ The paper used in this publication meets the requirements of the American National Standard for Information Sciences—Permanence of Paper for Printed Library Materials, ANSI Z39.48-1992.

www.bucknell.universitypress.org

Distributed worldwide by Rutgers University Press

Manufactured in the United States of America

CONTENTS

ILLUSTRATIONS

Romantic Automata

INTRODUCTION

Michael Demson and Christopher R. Clason

The editors of this volume share in the excitement surrounding the bicentennial of Mary Shelley's first edition of *Frankenstein* (1818) that has inspired a surge in books, exhibits, lectures, performances, and conferences on automata, assembled bodies, and artificial intelligence. The editors of *Frankenstein: Annotated for Scientists, Engineers, and Creators of All Kinds* (2017), it too a product of this excitement, aptly offer that Shelley's novel's "fusion of science, ethics, and literary expression provides an opportunity both to reflect on how science is framed and understood by the public and to contextualize new scientific and technological innovations, especially in the era of synthetic biology, genome editing, robotics, machine learning, and regenerative medicine" (xii). Their point is well taken: because of their attention to constructions and definitions of "life," Romantic texts are enjoying a renewed relevance in the first quarter of the twenty-first century, when once again we find ourselves trying to define what it means to be "human." Moreover, the critical turn in recent years to various posthumanisms—the idea that no definition of the "human" is stable—has reanimated interest in those Romantic texts that challenge cultural understandings of "life," human or otherwise. Looking beyond *Frankenstein* to a wide range of Romantic texts, this volume aims to demonstrate the manifold ways in which posthumanism has come to bear on the contemporary study of Romanticism, and what a return to Romanticism has to offer to current critical discussions.

Posthumanism is a compelling set of theoretical concerns that span across disciplines as diverse as environmental studies, philosophy, digital sciences, and literary studies (to add to an ever-growing list). Its central assertion, simply put, is that if we hope to understand our dependencies upon the world in which we live, how our lives participate within larger (and smaller) systems, we must recognize our deep-seated psychological, philosophical, political, and cultural inclinations to ignore or deny those very dependencies. Nevertheless, those dependencies are our way in to an understanding of how we live not apart from but enmeshed in the world. Richard Grusin, who prefers the term "the nonhuman turn" to "posthumanism" because he believes the latter misleadingly implies some teleology or historical development (as if we were no longer 'human'), identifies the salient feature of the critical trend:

> The nonhuman turn is to account for the simultaneous or overlapping emergence of a number of different theoretical or critical "turns" [in recent years]—for example, the ontological, network, neurological, affective, digital, ecological, or evolutionary. . . . Each of these different elements of the nonhuman turn derive from theoretical movements that argue (in one way or another) against human exceptionalism, expressed most often in the form of conceptual or rhetorical dualisms that separate the human and the nonhuman—variously conceived of as animals, plants, organisms, climatic systems, technologies, or ecosystems. (ix–x)

Indeed, we have held for a long time and with great tenacity onto anthropocentric convictions of "human exceptionalism," maintaining a faith in the independence of human agency, a self-defining consciousness, and an assumption of individual transcendence. That in turn has done more to obscure than bring into focus our shared, entangled, and integrated life on Earth with everything nonhuman. Posthumanism, by contrast, calls into question the conventional assumptions about human autonomy, rationality, and agency, so dear to thinkers since the Enlightenment. In doing so, posthumanism has opened up a wide range of questions in the twenty-first century that call for the reexamining of representations of life, bodies, and volition, as the essays in this volume all do.

Why return to Romanticism? The roots of that "Humanism," from which posthumanism strives to be distinct, admittedly stretch back much further in history than the Romantic era—yet it was the Romantics who first collectively perceived a threat posed to the security of our notions of the "human" in particular, and to an anthropocentric view of the world more broadly. They recognized this threat in the figure of the automaton. If the Romantic era is remembered for its celebration of human transcendence taken even to mythical levels, such as William Blake's *Universal Man* or Georg Wilhelm Friedrich Hegel's *Geist*, equally memorable are the frenzied, reactionary denunciations of nonhuman beings that recur throughout Romantic literature. In fact, the two strains are intimately twined, as Mary Shelley well perceived: the Romantic fascination with human creative genius, and the repulsion from the nonhuman and base materialism of the body. There is no better example than the confrontation between Dr. Frankenstein and his creature. This volume collectively seeks to argue that through their spirited responses to automata, Romantics had their own "nonhuman turn," to borrow Grusin's term, roughly two hundred years ago—Romantic authors raised deeply disconcerting questions about the constitution of life that many, confounded by their own inquiries, soon came to regret, just as Dr. Frankenstein sought to destroy what he had raised through his brilliant ingenuity. Fortunately for us, in doing so, their works speak with remarkable relevance to our current critical discussions.

Though he largely limits his claims to Paris in the 1830s and 1840s, John Tresch has argued in *The Romantic Machine: Utopian Science and Technology after Napoleon* that scientific thought of the Romantic era, including attitudes toward machinery and technology, was distinct from that of the earlier eighteenth and the later nineteenth century. He discusses a craze for scientific, and quasi-scientific, exhibitions in which inventions of various sorts, from new musical instruments through optical scopes to panoramas, were used to produce novel effects upon audiences who would bear witness to the transformative power of human ingenuity. Moreover, this technology period presented the natural world in a manner that "confounded familiar oppositions," such as subject and object or mind and matter, reconfiguring notions of knowledge, social relations, and the natural world at the same time as "fueling revolutionary desire for a future state of wholeness" promised "by the new powers of industry" (14–15). Not only did Romantics break from the epistemological and metaphysical paradigms of the Enlightenment, with its limited notions of machinery as mere clockwork, but also they entertained utopian notions of the transformative potentiality of human technology at the dawn of industrialization—notions that would form "the background both for a radicalized republicanism and for the birth of modern socialism" of the later nineteenth century: "New instruments and machines were theorized as extensions of human senses and intentionality, as fluid mediators between mind and world, and as the ligaments of society; they appeared as transformative, even sublime devices" (5–6). Because they "posed a direct challenge to the dualism of matter and spirit," Tresch suggests, the automata of the era exemplified this Romantic science (153).

The great European automata-makers of the early to mid-eighteenth century—Jacques de Vaucanson, Wolfgang von Kempelen, Pierre Jaquet-Droz (whose writing automaton, *L'écrivain*, is featured on the cover of this volume)—produced works that were astounding in their mechanical ingenuity, and their exhibitions inspired popular wonder and delight. These were, according to Jessica Riskin in *The Restless Clock: A History of the Centuries-Long Argument over What Makes Living Things Tick*, "machines that (actually or apparently) ate, shat, bled, breathed, cavorted, walked, talked, swam, made music, drew, wrote, and played an almost unbeatable game of chess" (123). And the great automata inspired the mass production of simple automata. In her *Letters Written in France*, Helen Marie Williams speaks of a snuff box she purchased in Revolutionary Paris: "You touch a spring, open the lid of the snuff-box, and the Abbe jumps up, and occasions much surprise and merriment" (86); Nikolai Karamzin mentions the same toy in his

Travels from Moscow through Prussia, Germany, Switzerland, France, and England, and Mary Wollstonecraft carries another, with a pop-up mouse, in her travels across Sweden, Norway, and Denmark (10). Automata were celebrated as feats of human science, reason, and ingenuity, but this wonder and delight soon gave way to anxieties about the mechanical simulation or duplication of life in the later part of the century, as identifying precisely what the term "human" designated proved elusive.

In fact, by the mid-century, *homo sapiens* had already substantially lost its unique position at the very center of creation, while in the rationalist view of life on the planet the species was becoming connected ever more closely to others in the animal kingdom. The inner workings of human mind and spirit, which for centuries had exempted humanity from the rest of nature, were viewed increasingly as mechanisms subject to the same physical laws as those governing *fauna* and *flora*, and perhaps even the machine. Riskin identifies the start of modern thought on automata with René Descartes, who in abstracting the mind from the body effectively rendered both the body and, by extension the world, a machine upon which the mind, and God, operated (44–76). (Many Romantics took up this notion of the world as machine, but played upon this metaphor in complex ways, as Christine Weiler's essay explores.) Descartes had already suggested as early as 1637 that animals were merely automata and differed from *homo sapiens* merely by virtue of a soul, which humans possessed (and which Descartes located) in the pineal gland.[1] Some of the more materialistic French *philosophes*, such as Julian Offray de La Mettrie, had suggested that human beings were essentially "organic automata," and that perception, thought, and behavior were no more than effects of causes that were, essentially, chemical and mechanical in nature.[2] By 1810, Heinrich von Kleist, writing of the superiority of marionettes to human dancers with his characteristic irony and deliberate provocation, argued that, "Where grace is concerned, it is impossible for man to come anywhere near a puppet" (7). He attributed the affectation of human dancers, their lack of grace, to our postlapsarian self-consciousness that alienates the dancer from the dance, whereas the "lifeless, pure pendulums, governed only by the law of gravity" (6)—or the innocent boy (8) or wild bear (10) who equally lack self-awareness—have none of that consciousness that necessarily "disrupts natural grace" (8): "Grace appears most purely in that human form which either has no consciousness or an infinite consciousness," he factiously concludes, "that is, in the puppet or in the god" (12). Of course, many of the literary accounts of automata strain credulity, as Riskin suggests (127–128) and as Wendy Nielsen's chapter studies here in depth.

Irony aside, by the end of the eighteenth century, a deep dread of puppets and automata, and the machinery that propels them, had surfaced in British and

European Romantic literature, as it would later in America and Russia. These "mere images of living death, or inanimate life" (as E.T.A. Hoffmann's character Ferdinand famously dubbed them in the tale "Automata" [1814])[3] prompted a cultural paranoia and repulsion to mechanization and contrivance: such science and engineering—regardless of whether it was literary, theatrical, biological, imperial, or industrial—threatened the spiritual nature of life, the source of true compassion in human society. As early as 1762, Rousseau was using "automata" pejoratively, arguing that the effect of modern society on a child was to render him—Émile—"a perfect imbecile, an automaton," and that only a strict adherence to "natural education," away from society, could prevent this fate (*Émile* 61, 91). Rousseau's influence was ubiquitous in European and British Romanticism. As they proclaimed political independence, philosophical transcendence, and spiritual autonomy, Romantics also decried the strings and cables, cogs and gears, hidden crews and inhuman machinations, micro and macro systems, at work backstage and in the dark, within and without, that threatened human autonomy. They imagined hidden conductors like Shakespeare's Prospero orchestrating social plots, puppeteers like Faust manipulating people's motivations, hidden dwarves working unseen clockwork to stage awful spectacles—all too often in order to influence people, to win profits, or to gain political advancement. Romantic villainy was measured by the ingenuity of villains, the complexity of and control over their machinations and artful designs, rather than the baseness or immorality of particular crimes; modern villainy became synonymous with the hubris of steering larger-than-human machinery, be it immaterial social forces, such as public opinion or financial markets, or actual technical machinery, such as the canon (by which Napoleon won his early notoriety). Might such malevolent agents with their soulless contrivances displace our sense of what is truly human, or perhaps worse, might the labors of machines duplicate and thereby replace our own, rendering us obsolete? Gothic horrors and fantastic tales proliferated.

And these fears were not echoed in works of literary fancy alone—this dark side of Romanticism runs through the great political writings from the revolutionary era: Jefferson's celebrated "self-evident" "truths" from the American *Declaration of Independence* (1776)—"that all men are created equal"—begs the questions of which humans are covered by the designation "all men," and if not equal then what is the status of those not covered by that designation? Radicals, early feminists, and abolitionists of the Romantic era recognized that the working classes, women, and slaves were systematically silenced, brutalized or objectified, manipulated as tools, and, of course, discredited as human beings. They argued fiercely that these nefarious socio-economic systems affected everyone, even those who denied any involvement. For example, in the "Debate on Mr. Wilberforce's Motion for the Abolition of the Slave Trade" on April 2, 1792, William Wilberforce stressed that the violence intrinsic to the institution rendered both slaves and slavers into beings less than human, that "in every point of view, [slaves] were regarded and treated as animals of a distinct species from man" (163) while "those inhuman instruments of oppression whipped them like cattle, not supposing them moral agents, capable of reflection or resistance" (164).[4] Such rhetoric was redoubled by the British radicals of the early 1790s, who extended the condemnation of the institution of slavery to the whole institution of British government—by 1795, Samuel Taylor Coleridge was publically condemning the government's use of "uncouth and unbrained automata," or hired thugs, to intimidate him and other radicals into silence.[5] In speaking of Crown Prince Frederick, Mary Wollstonecraft uses similar tropes:

> I cannot describe to you the effect it had on me to see this puppet of a monarch moved by the strings which count Bernstorff holds fast; sit, with vacant eye, erect, receiving the homage of merely a machine of state, to subscribe the name of a king to the acts of the government, which, to avoid danger have no value, unless countersigned by the Prince Royal; for he is allowed to be absolutely an idiot, excepting that now and then an observation, or trick, escapes him, which looks more like madness than imbecility. . . . What a face is life! (103)

The same language reappears is Percy Shelley's radical tract, "A Philosophical View of Reform":

> It is asked, how shall this [reforms] be accomplished, in defiance of and in opposition to the constituted authorities of the Nation, they who possess whether with or without its consent the command of a standing army and of a legion of spies and police officers and hold all the strings of that complicated mechanism with which the hopes and fears of men are moved like puppets? (667)

The hyperbolic rhetoric of the British radicals of the 1790s, taken up by Shelley in later decades, aimed at identifying and deriding the machinery of government and its complementary ideology as so unnatural that nothing about it could be humane—politicians, and those under their influence, were literally not human. At the climax of Shelley's gothic play, *The Cenci*, Cardinal Camillo describes the villainous Pope:

> The Pope is stern; not to be moved or bent.
> He looked as calm and keen as is the engine

> Which tortures and which kills, exempt itself
> From aught that it inflicts; a marble form,
> A rite, a law, a custom: not a man.
> He frowned as if to frown had been the trick
> Of his machinery . . . (V, iv)

Here "engine" and "machinery" describe that which is "not a man" and the actions of a nonhuman agent, so cruel and so without compassion or sympathy, that the agent cannot be deemed human.

For the first time, literary works focused not as much on the objective phenomenon of the automaton per se, but rather on the subject's psychological reaction to viewing such clever imitations of life. Tresch writes, "The romantic machine did not stand alone; it involved the active participation of the observer and articulated a spontaneous, living, and constantly developing nature; it produced aesthetic effects and emotional states . . . —it was imbued with the aesthetics and the affects of the organic, the vital, and even the transcendent" (12). Sometimes the subject remained aware of the indications of lifelessness—cold and dull eyes and stiff, repetitive and perfectly rhythmic motions, disproportionate body parts, machine noises, skin cold to the touch—that constantly reminded the viewer of the automaton's essentially mechanical nature. For example, the British caricaturists George Cruikshank and William Heath both produced satirical prints in 1828 of colossal personifications of unrestrained industrial technology, which no one would mistake for humans, wreaking havoc on humanity: "London going out of town—or, the March of Bricks & Mortar" and "March of Intellect," respectively.[6] At other times, the automaton was harder to detect, or the viewer was unwilling, often for amorous reasons, to see it as artificial. In these instances, Romantics frequently returned to the myth of Pygmalion, the sculptor who fell in love with his own sculpture, only to be undone by its coldness. The perplexity of sensual contradictions, the insecurity and embarrassment of doubt, the perverse fascination with the abnormal or monstrous, all drove Pygmalion toward insanity—were it not for Aphrodite's kind, animating touch that brought Galatea to life. Such plots became commonplace in Romantic literature; the lack or loss of humanity became a defining counter-theme of the period, with its insistence—sometimes belligerent, sometimes ironic—on human genius, moral and political independence, and defiant action that sought to prove, if nothing else, human autonomy. This is the tension at the heart of Nathaniel Hawthorne's "The Birth-mark," in which a genius scientist kills his "imperfect" person in his attempt to render her perfect by removing her flaw.

Whether blatant or subtle, uneasiness about soulless machinery is indeed palpable throughout the broad literary landscape. E.T.A. Hoffmann's German clockworks, Mary Shelley's Swiss animated corpse, Nathaniel Hawthorne's American dandy scarecrow in "Feathertop" or poisonous daughter in "Rappaccini's Daughter," all incorporated the uncanny; they expressed an essential distrust of such mechanisms, engineered to seem conscious and animated, moving under their own volition and simulating, if not mocking, the appearance of living, autonomous and sentient beings. It is perhaps this distrust that inspires "the overriding desire of most little brats," as Charles Baudelaire poignantly remarks, "to get at and *see the soul* of their toys":

> When this desire has planted itself in the child's cerebral marrow, it fills his fingers and nails with an extraordinary agility and strength. He twists and turns the toy, scratches it, shakes it, bangs it against the wall, hurls it on the ground. From time to time he forces it to continue its mechanical motions, sometimes in the opposite direction. Its marvelous life comes to a stop [. . .] finally he pries it open, for he is the stronger party. But *where is its soul*? This moment marks the beginnings of stupor and melancholy. (24)[7]

The recognition of one's own psychological projections as such does not inspire so much as it depresses our own thoughts because it reminds us of the materiality of life, our confinement to body. In doing so, as Romanticism repeatedly attests, such reflections inspire the desire to destroy what has been created, to lash out with violence, and to turn against oneself.[8]

Less fantastic but perhaps even more prevalent was the new use of "automata" as a term of disparagement for "soulless people"—the easily incited mobs of William Godwin's or Victor Hugo's novels, the mindless women of Baudelaire's poetry or the targets of the early British feminist satires of Mary Hays and Sydney Owenson, or even the bald-faced satire of Nikolai Gogol's *Dead Souls*. Romantics used the figure of "soulless people" in order to distinguish their own independence, but in doing so, they articulated a fear that their own subjectivity relied upon hidden machinery, systems, and apparatuses. As Romantic characters began to imagine themselves as transnational subjects in a rapidly globalizing world, they also saw dark, fearsome reflections racing before them, traveling alongside them, or chasing after them—of cosmopolitan *Doppelgänger*, mechanical foreigners, and other, seemingly sentient puppets, striving to supplant their human counterparts. Beyond fear and anxiety, British and European Romantics were composing in the early age of globalization and industrialization, and regardless of their own poetical or ideological pronouncements, these transformative processes affected their sensibilities. Indeed, seeing past Romantic expressions of anxiety and revulsion,

Tresch has noted the "close partnership between Romanticism and mechanization" that extended well beyond Enlightenment materialism and science, into "the supernatural undercurrents of the early industrial age" (87). British imperialism, for example, brought Sanskrit texts across the oceans from colonial India, upsetting and inspiring the British poet Robert Southey who (rather disingenuously) complained that Hinduism—"which of all false religions is the most monstrous in its fables"—could not "supply fit machinery for an English poem," in his preface to his very own Oriental epic, *The Curse of Kehama* (1810). Southey's ambivalence, both fascination and repulsion, in regards to what he saw as novel foreign religious contrivances is paradigmatic of the period: that which simulates life, or stages life, holds us enthralled.

It is Sigmund Freud who most notably studied this fear, as it was expressed in Hoffmann's tales, and whose 1919 essay "Über das Unheimliche" ("On the Uncanny")[9] has ever since provided a touchstone for scholars interested in Romanticism and automata. In this essay, Freud poses a question: what is the quality of an "uncanny" (*unheimlich*) phenomenon that differentiates it from the other kinds of experiences that inspire dread or terror in the psyche. Exploring the German term etymologically, he discovers that the root, *heimlich*, bears a rich span of meanings, and that it "becomes increasingly ambivalent" (134)—not only does it describe things which are "familiar" or "belonging to the home," but also that which is "secret" and "hidden from public view"—quoting the Grimms quoting Schelling, Freud finds a particularly fruitful nuance of meaning in the statement, "uncanny is what one calls everything that was meant to remain secret and hidden and has come into the open" (132). Thus, the term "finally merges with its antonym, *unheimlich*" (134). Then, as a case study, Freud provides a reading of Hoffmann's "The Sandman" ["Der Sandmann," from the collection *Nachtstücke*, 2 vols. 1816–1817], a tale in which the student protagonist, Nathanael, who has described in a letter a trauma he suffered as a child that involved a threat of losing his eyes to a wicked "Sandman," becomes amorously obsessed with an automaton created by a physicist-machinist named Spalanzani and his colleague, the latter of which bears the name of, and is identified with, Nathanael's nemesis (Coppelius/Coppola). In fact, it is this character to whom Freud links the major portion of the tale's uncanniness—the Sandman, who threatens to rob the child's eyes, evokes a castration fear in Nathanael parallel to that found in "dreams, fantasies, and myths" (140).

Thus, despite the remarkable turn of events that leads to Nathanael's love for an automaton, Freud insists that the major consideration here is the young man's fear of the loss of eyes and sight. However, Freud's discounting of the automaton's importance ignores aspects of uncanniness as he defines it—which include "animism, magic, sorcery, the omnipotence of thought, unintended repetition and the castration complex" (149). It is, in fact, when the boy hides behind a curtain and observes his father and Coppelius engaging in what we can only surmise to be an early attempt at automaton construction that the two adults discover him and drag him out onto their workbench, threatening him with the loss of his eyes and then manipulating him *as if he were an automaton*. When, near the conclusion of the tale, he later encounters Spalanzani and Coppelius in a tug-of-war, fighting for possession of the lifeless Olimpia, his beloved automaton, Nathanael realizes for the first time that she is only that; but now he also discovers that there are merely empty sockets where her eyes had been. Spalanzani tells Nathanael that Olimpia's eyes had been stolen from him when he was a child, and he throws them at the young man, sending him into a mad rage. Thus, the "uncanniness" of this scene relies not only on the eyes but also on the automaton—which, bearing Nathanael's eyes, becomes a *Doppelgänger*, the mechanical "double" to the young man whose trauma originally involved his being treated as if he were nothing more than a mechanical puppet whose eyes could be popped out of his head and re-appropriated for other uses. Thus, the threat to Nathanael is castration, but it is also expressed as a loss of identity, even a loss of humanity, through his encounter with the automaton.

While Freud's approach to our understanding of the Romantic fascination and horror of automata remains a touchstone essay, recent scholarship in posthumanism, postcolonialism, disability studies, postmodern feminism, eco-criticism, and radical Orientalism has guided the critical discourse on this topic in significant directions, as this collection aims to demonstrate. To that end, each of the essays in this collection challenges the Freudian paradigm by opening up new methodological approaches to understanding human interaction, interdependence, and interrelationships with technology that strives to simulate or to supplement organic life. The first essays of this collection, gathered under the subtitle "Exhibitions," keep close to the Romantic understanding of the automaton as primarily a theatrical spectacle, encountered as an artwork in a gallery.

In his essay "The Uncanny Valley: E.T.A. Hoffmann, Sigmund Freud, Masahiro Mori," Frederick Burwick moves deliberately beyond Freud's reading of Hoffmann's tales in "Das Unheimliche" (1919) to explore Masahiro Mori's thesis from "The Uncanny Valley" (1970) that the revulsion associated with a puppet's approximation of human appearance requires that the puppeteer remain hidden from view. When this is not the case, when the audience sees both puppet and puppeteer, the effect is entirely different. Burwick revisits Hoffmann's tales to identify those moments in which characters see puppeteers working their contrivances and demonstrates that these moments of cooperative movements are not anxiety producing, as Freud would suggest, but moments of levity or gratification, such as Marie with her nutcracker in "Nußknacker und Mausekönig."

Following Burwick's essay, Ashley Shams explores theatrical reinterpretations of Hoffmann's automata in "The (Re-)Winding of Hoffmann's Automata: From Offenbach's 1881 Opera to Powell and Pressburger's 1951 Film." In adapting Jacques Offenbach's *Les contes d'Hoffmann* to their film, she argues, British filmmakers Powell and Pressburger chose to redefine the role of Olimpia from a singing automaton to one who also dances ballet. In large part, Shams argues, this change was motivated by a sense that the dancing (human) doll in the film captures the essence of automata in a more immediate way. Elements such as the lip-synching of the actor/dancers to the pre-recorded sound of off-stage singers, the set designs involving the use of trompe l'oeil, the mixing of "real" and human marionettes, and other similar cinematic techniques all demonstrate how changes in medium (stage opera to film) and genre (opera to ballet) amplify the effect of the automaton in Hoffmann's tale and heighten the sense of the uncanny, allowing the viewer to be suspended between the real and unreal of the cinematic experience. Indeed, the film's emphasis on performance and its use of cinematic trickery echo the mechanical trickery and performative aspects of Hoffmann's original automata.

In the final essay of "Exhibitions," Peter Erickson explores the end of the brief career of the French Romantic painter, Théodore Géricault, when he became enthusiastic for the invention of lithography, the latest advance in the mechanical reproduction of images for a mass audience. Erickson discovers that a surprising number of Géricault's lithographs depict wounded veterans or disabled panhandlers as they make use of crutches, canes, or prosthetic limbs. Moreover, Géricault employs these prosthetic limbs and crutches to create unorthodox and often even uncanny compositions, in which wounded or disabled bodies lean on one another for support. Géricault, Erickson continues, depicts prosthetics in order to generate our emphatic conviction in the unity of the *tableau*, that is, our conviction in the unity of the work of art. One can indeed trace, in Géricault's sketches, how he experimented with precisely which wounds he should inflict in order to render the bodies depicted in his lithographs ever more dependent on each other. As a result, these lithographs represent nothing less than a radical de-privileging of the human body in the early nineteenth century as the fundamental, indivisible unit of history painting.

The essays of the second section of this collection, gathered under the subtitle "Figures," return to Romantic representations of automata, and more specifically to the recurrent language and figures of speech, including irony, deployed in these representations. In her essay, "Romantic Tales of Pseudo-Automata: Jacques de

Vaucanson and the Chess-Playing Turk in Literature and Culture," Wendy Nielsen surveys literary depictions of automata to argue that tales about pseudo-automata and androids function as meta-narratives about the potential for technology to change global economies. She discusses how reflections on popular stories about Descartes, Vaucanson's duck, or von Kempelen's "Chess-Playing Turk," for example, led such divers writers as Hoffmann, Edgar Allan Poe, and Walter Benjamin to become skeptical about the purposive character of transcendental natural philosophy, on the one hand, and global capitalism and the conditions of labor on the other. In other words, the authors were not concerned that the automaton was fake but that the figure of the automaton distorted the perception of the rapidly changing conditions of labor. Perhaps one may discern a connection here between this point and the fact that Amazon's "marketplace" for Human Intelligence Tasks is called "The Mechanical Turk" (MTurk),[10] after von Kempelen's "Chess-Playing Turk."

Similarly, in "Rattled Women, Shaken Toys: Wollstonecraft, Baudelaire, and the Musical Lady," Erin Goss considers the figure of the eighteenth- and nineteenth-century automaton as it comes not only to embody sexual difference but also to provide license for forms of gendered violence. She tracks the association of women with automation in fiction and conduct manuals of the period so as to underscore the Romantic notion that "femininity" is the desired result of an education in manners. That is, women's education (so ran the conventional line) should consist of the training in habit through repetition; their education is an inculcation of manners rather than morals. And yet, if they become too mechanical, too automatic, they are berated for their automation and denigrated as no longer sufficiently human. As a result, sexual difference often appears as a differential relation to automation: one can embody the mindlessness of habit, or one can assert one's will. To be too habitual is to have been taught to be a woman, to show oneself to be, as Mary Wollstonecraft puts it in *A Vindication of the Rights of Woman*, "created to be the toy of man"; to act voluntarily is to be masculine. Moreover, the automaton not only becomes a figure for the excessively mechanical qualities of eighteenth-century femininity but also provides license for a fantasy of violence against both the mechanical and the feminine.

Arguing along similar theoretical lines, Kate Singer contends in "Automatic for All: Mary Shelley's Posthuman Passion" that Shelley's short fiction recurrently asks us to think deeply about the intersections among all sorts of beings—humans, animals, and nonhumans. Singer maintains that Shelley attends to questions about soulfulness, volition, and embodiment, in recurrent moments that experiment with a mutual interchange of passion, materiality, and technology—that is, human entanglement with the technical, economic, and emotive, in a material passion that moves contingently through various bodies even as it transgresses death. The alchemical concoction, the body-switching pact, the science of reanimation, and boundary-crossing passions all manifest, at their core, an undoing of the boundaries of embodiment that would separate human from nonhuman, remaking those bodies while moving them. Shelley's tales, in this way, explore a recursive, material affect that crosses and remakes ontological boundaries between human, animal, and thing by pointedly reconsidering the categories of autonomy, volition, and soulfulness, nonhuman automaticity, and animal instinct.

The last essay in "Figures" explores what might be taken for an inversion of figure of the Romantic automaton. In "'A little earthly idol to contract your ideas': Global Hermeneutics in Phebe Gibbes's *Zoriada, or, Village Annals* (1786)," Kathryn Freeman explores how this recently rediscovered oriental tale casts a complicating light on the topic of the late eighteenth-century revolt against Enlightenment mechanization and science. The novel centers on the incongruity of the pseudonymous "Zoriada," a young Indian woman suddenly appearing in a rural English village. Her mysterious identity gives rise to and ultimately rises above the villagers' "speculation," Gibbes's ironic term leveling this microcosmic society's spectrum of responses, from gossip to scientific deduction. In the hermeneutical quest to find meaning in the text, the reader, too, is implicated in the projection of identity onto Zoriada, who remains a cypher until the novel's end when she herself decides to reveal her subjectivity. The novel thus helps re-think the traditional paradigm of a dualistic shift between Romantic sensibility and Enlightenment reason by positing a nondual relationship to knowledge represented by the exotic Zoriada, both symbol and teacher of an epistemology that eludes western binaries, including those of gender and selfhood. Zoriada's own manuscript, discovered in the novel, remains untranslated, just as the reader, Freeman concludes, is left to speculate upon what is knowable through the limited paradigm of western epistemology.

The final grouping of essays, "Organisms," moves away from questions of humanity and subjectivity to explore issues concerning "the organic," another figure often defined antithetically in relation to machinery and automation. In doing so, all three of these essays challenge traditional understandings of the metaphor "organic" by exploring Romantic authors who pushed this term beyond conventional recognition. In the first essay of this section, "Schelling's Uncanny Organism," Stefani Engelstein takes her cue from Georges Canguilhem's observation in the 1950s that there was a kind of suspended oscillation between mechanism and vitalism in the emerging biological discourse around 1800—a dialectic of the "organism." She goes farther, however, arguing that the emerging concept of the organism conjured ontology out of conundrum, crystallizing as a name for all living beings a word that had previously, like "mechanism" and "vitalism," specified a particular but contested quality of organized systems. Schelling, for example, attempted to achieve a unified theory of nature that could encompass both will and matter through a concept he described simultaneously as a world-soul and a universal organism. By binding the universe together in ways that emphasized reciprocal interactivity, balance, growth, regeneration, and dissolution, but to differing degrees in different instantiations, Schelling hoped to eliminate the untenable uniqueness of living things in a material world and to construct a place for human will. If the early modern image of the world as a machine designed by God with special dispensations for human, willed action had slowly metamorphosed into a machinery of nature in which humans themselves were trapped as puppets, the organic view both of the universe and of the individual living being did not free humans from constraint but rather entangled them in a complex expressive system of drives and purposes.

The second essay in "Organisms," Lenora Hanson's "'It . . . lives by dying': S. T. Coleridge's Non-Vital Life and Colonial 'Necoral-politics,'" closely examines two enigmatic statements made by Coleridge about living bodies, or more specifically, about the curious status of coral as a vital organism. She argues that, for Coleridge, coral manifests a profound integration of mechanical and vital life, its growth dependent upon its perpetual death. In other words, coral confounds our conventional opposition of the living and the dead by presenting a figure in which both were entirely integrated. This figure of coral, Hanson then proposes, affects on a much larger level all of Coleridge's geopolitical thought. No longer is a "society," "colonialism," or "rebellions," either a megamachine or a living thing, either living or dead. Instead, Coleridge's "Theory of Life" helps us to envision one necropolitical, global assemblage that integrates the living and the dead.

Such an integration of life and death takes place similarly on (and "within") the land, and bears a special connection to automation: mining. Christina Weiler argues in the final essay of the collection, "The Metaphysical Machinery of Mining in Novalis's Works," that Friedrich von Hardenberg, better known as Novalis, challenges the metaphor of "organic." She tracks how Novalis characterizes the machinery of small-scale artisanal mining and how his characterizations elicit epistemological and phenomenological insights into the depths of the world and our dependencies upon it. While Novalis's professional notes from his time as a saline assessor list material objects and costs, his poetic works are much more speculative about mining. That is, the mine presents a liminal space that enables the human being to see the organic harmony of nature and self over vast extents of time. The wooden beams, on which the miners descend into the mountain, and the baskets, with which the miners move the stones, are highlighted as organic parts in the mining machinery. Even the metal, from which the hammers and other mining instruments are made, seems to grow plantlike out of the ground—the organic and the mechanic do not stand in opposition to each other but rather represent a larger, organic harmony that encompasses all of nature, including the human being. The images of Heinrich von Ofterdingen's life, collected in a mysterious book he finds, mirror the cyphers of the fossils and metals embedded in the mountain. What enables the hermeneutic process of deciphering these writings of nature in the mountain is a mining instrument: the old miner's lamp. The emphasis on the organic and on light and fire in the description of the mining machinery in *Heinrich von Ofterdingen* reflects metaphorically the epistemological and phenomenological insights that Novalis envisioned through his aesthetic mining organon.

NOTES

1. Minsoo Kang, *Sublime Dreams of Living Machines: The Automaton in the European Imagination* (Cambridge, MA: Harvard University Press, 2011), 117.
2. Kang, *Sublime Dreams*, 130.
3. E.T.A. Hoffmann, "Automata," in *The Best Tales of Hoffmann*, trans. E. F. Bleiler (New York: Dover, 1967), 81 ["diese wahren Standbilder eines lebendigen Todes oder eines toten Lebens"; see E.T.A. Hoffmann, *Sämtliche Werke*, vol. 4, *Die Serapions-Brüder*, ed. Wulf Segebrecht w/collab. Ursula Segebrecht (Frankfurt am Main: Deutscher Klassiker Verlag, 2001), 399].
4. "Debate on Mr. Wilberforce's Motion for the Abolition of the Slave Trade," *The Parliamentary History of England from the Earliest Period to the Year 1803*, vol. 29 (London: Longman, Hurst, Rees, Orme & Brown . . . , 1817), col. 1063–1064.
5. To put this quote in the context of radical culture in the 1790s, see Stephen Gill, *William Wordsworth: A Life* (New York: Oxford University Press, 1989), 93, and Adam Sisman's *The Friendship: Wordsworth and Coleridge* (New York: Viking, 2007), 99.
6. Cruikshank's caricature is at the British Museum and can be found reproduced at http://www.britishmuseum.org/research/collection_online/collection_object_details.aspx?objectId=1663147&partId=1 and Heath's is at the British library, https://www.bl.uk/collection-items/march-of-the-intellect (last consulted March 30, 2019).
7. Charles Baudelaire, "The Philosophy of Toys," in *Essays on Dolls*, trans. Idris Parry and Paul Keegan (New York: Syrens/Penguin, 1994), 15–25, 24. ["Morale du joujou," in *L'art romantique*, 4th ed., ed. Calmann Lévy, Oeuvres complètes, vol. 3 (Paris: Michel Lévy Frères, 1879), 139–149, 148, accessed March 31, 2019, https://babel.hathitrust.org/cgi/pt?id=pst.000018643770; "La plupart des marmots veulent surtout *voir l'âme*, les uns au bout de quelque temps d'exercice, les autres *tout de suite*. C'est la plus ou moins rapide invasion de ce désir qui fait la plus ou moins grande longévité du joujou. Je ne me sens pas le courage de blâmer cette manie enfantine: c'est une première tendance métaphysique. Quand ce désir s'est fiché dans la moelle cérébrale de l'enfant, il remplit ses doigts et ses ongles d'une agilité et d'une force singulières. L'enfant tourne, retourne son joujou, il le gratte, le secoue, le cogne contre les murs, le jette par terre. De temps en temps il lui fait recommencer ses mouvements mécaniques, quelquefois en sens inverse. La vie merveilleuse s'arrête . . . enfin il l'entr'ouve, il est le plus fort. Mais *où est l'âme*? C'est ici que commencent l'hébétement et la tristesse."]

8. Erin Goss's contribution to this volume examines this passage closely and shows how in Baudelaire's analysis of children's reaction to dolls, they not only discover deeper, disturbing new feelings, but also the "parameters of lived sexual difference" as well as a definition of "the relation between the sexes by violence."
9. Sigmund Freud, "The Uncanny," in *The Uncanny*, trans. by David McClintock, with an intro. by Hugh Haughton (New York, Toronto, et al.: Penguin, 2003), 123–162.
10. The Amazon "MTurk" can be accessed at https://www.mturk.com/ (last consulted January 13, 2018).

Part One

EXHIBITIONS

1

THE UNCANNY VALLEY

E.T.A. Hoffmann, Sigmund Freud, Masahiro Mori

Frederick Burwick

In this essay, I examine conceptual intersections among texts by three authors: Hoffmann's "Automata" ["Die Automate," 1814][1] and "The Sandman" ["Der Sandmann," 1817],[2] Freud's "The Uncanny" ["Das Unheimliche," 1919],[3] and Masahiro Mori's "The Uncanny Valley" (1970).[4] I would guess that at least half of those who have taught or written about "The Sandman" have introduced Freud's "The Uncanny" for critical leverage. The purpose of Freud's essay is to discuss the consequences of projecting personal fears and desires onto exterior persons, objects, or events. Mori argues in his essay that lifelike robotic figures may become increasingly attractive and fascinating up to a point at which their approximation of human appearance prompts an awareness of the non-human qualities which turn the attraction into a revulsion akin to that felt in response to the animated dead. I argue that all three of these authors implicate a doubleness of response, in which attraction is turned to repulsion, *canny* is turned to *uncanny*, *heimlich* to *unheimlich*. I further emphasize that the doubleness is inherent and persistent.

Freud, we recall, cited Hoffmann's tale as his prime example of the *uncanny* or *das Unheimliche*. Nathanael's experience of the uncanny arose from his aberrant projection of his own psychological responses onto external circumstances. One instance was his fear of Coppelius/Coppola; another was his desire for Olimpia, whom he first perceived as the daughter of his professor, but then discovered her to be his professor's mechanical wind-up automaton. But his desire for Olimpia was not unaccompanied by his sense of something uncanny in her nature.

Freud discusses "The Sandman" in the second part of his three-part essay. The first part is devoted to an etymological and semantic construction of the meanings of *unheimlich*, especially in relation to its counterpart *heimlich*. Tracing in other languages the word pairs that closely resemble the relationship between *heimlich* and *unheimlich*, Freud suggests *canny* and *uncanny*. The fit is so close that

English translations of Freud's essay encounter little difficulty with the substitution. The English word *canny* is without relationship to *home*, and there is nothing disconcerting in its cognate kinship with to *ken* and to *know*. However, a *canny* person as a *cunning* person reveals the word's darker side, which becomes even darker with the negative prefix *un*. Similarly, something darker is lurking in the German *heimlich* which conjures the familiarity and comfort of home, but also may hint at the things that we keep hidden at home, the skeletons in our closets. A person who "heimlich gegen seine Feinde handel[t]" [acts *heimlich* against his enemies] is not treating them with down-home hospitality but with deceit and treachery. "Die heimliche Kunst" [the secret art] refers to sorcery rather than to unexhibited paintings in the artist's studio.

Mistakenly claiming to cite from Ludwig Tieck's translation of *Cymbeline*, 3.2, Freud may have thought of Imogen's praise of Leonatus for his power to read the stars. Instead Freud quotes from *1. Henry IV*. 3.1. This is the scene that opens with Hotspur taunting Owen Glendower for his pretension of supernatural powers, declaring that he "can call spirits from the vasty deep" (3.1.53). Mortimer then defends Glendower, saying that he is "Exceedingly well read, and profited / In strange concealments" [Ganz ungemein belesen, und erfahren / In seltnen Heimlichkeiten] (3.1.166–167). To capture the sense of the secret arts Tieck has translated "strange concealments" as "seltnen Heimlichkeiten." As Freud observes, the meaning overlaps with its supposed negation:

> Also heimlich ist ein Wort, das seine Bedeutung nach einer Ambivalenz hin entwickelt, bis es endlich mit seinem Gegensatz unheimlich zusammenfällt. *Unheimlich* ist irgendwie eine Art von *heimlich*.[5]

The same shared reference occurs in the English translation:

> [Thus *canny* is a word, the meaning of which has developed with an ambivalence, until it finally fell together with its opposite, *uncanny*. Somehow *uncanny* is a species of *canny* (my translation).]

The uncanny derives its terror not from something externally alien or unknown but from something "strangely familiar." The question thus arises whether the shared connotative range, attraction and repulsion, is consecutive or simultaneous.

Three years before "The Sandman" was published in the *Nachtstücke* (1817), Hoffmann's "Automata" appeared in *Die Zeitung für die elegante Welt* (14. Jg. Nr. 68–75 [Leipzig, 1814]). In his description of Olimpia playing the piano and dancing, Hoffmann draws upon actual figures of clock-work ingenuity that were being exhibited in Europe in the eighteenth and nineteenth centuries. Hoffmann created a character, Professor X, whose inventions followed those of Jacques de Vaucanson, whose *Flute Player* and *Tambourine Player* were first exhibited in 1737,[6] Wolfgang von Kempelen's *Chess-Playing and Fortune-Telling Turk* in 1770, and Pierre Jaquet-Droz's *The Scribe*, *The Musician*, and *The Draughtsman* between 1768 and 1774.[7] Professor X's mechanical and musical genius is the major topic of discussion between Ferdinand and Ludwig who visit his public showrooms and his private cabinet. They question the extent of possible interaction between human and mechanical instruments or replica, whether that interaction might be enhanced by a sympathetic or psychic rapport, or significantly altered if that mechanical instrument were also a musical instrument. After the tour of the Professor's studio, Ferdinand affirms:

> We have in fact seen remarkable mechanical works of art, even in regard to music. The flute player is apparently Vaucanson's famous machine, and the same mechanism utilized for the finger movements of the female figure, who brought such harmonious tones from her instrument; the accord among the machines is wonderful.
>
> [Wir (haben) aber in der Tat merkwürdige mechanische Kunstwerke gesehen; auch in musikalischer Hinsicht! Der Flötenbläser ist offenbar die berühmte Vaucansonsche Maschine, und derselbe Mechanismus rücksichtlich der Fingerbewegung auch bei der weiblichen Figur angewendet, die auf ihrem Instrumente recht wohllautende Töne hervorbringt; die Verbindung der Maschinen ist wunderbar.][8]

In contrast to the more musical and imaginative Ferdinand, Ludwig seems to accept a version of the dualism espoused by René Descartes and forwarded by Julien Offray de La Mettrie's *Man a Machine* [*L'homme machine*, 1748], crediting the emergence of the mind as the haunting "ghost in the machine."[9] Inventors of the automata sought to conjure the illusion of the haunting presence. In accord with that mechanist account of bodily behavior, Jacques de Vaucanson's mechanical duck (1739) displayed processes mimicking animal life. His clock-work duck could eat grain then defecate. The duck poop was fake, pre-prepared to make it seem as if mechanical functions could replicate those of living organisms. Vaucanson's plans for androids to serve dinner and clear the tables were welcomed by most, but were denounced by one bureaucrat who declared the blasphemy of the endeavor and ordered Vaucanson's workshop to be destroyed.[10]

The closest actual counterpart to "the speaking Turk" [der redende Türke] of Hoffmann's tale was the *Chess-Playing Turk* first exhibited by Wolfgang von Kempelen.[11] Its animation and game-skill depended on an actual chess player concealed in the base of the automaton. The fortune-telling apparatus depended on hidden tubes through which a spectator's question could be answered. The clock-work Turk of Hoffmann's tale is presented as if capable of liminal life and oracular insights. In dialogue with Ferdinand, Ludwig denounces the sham of all such animated creatures as "these true statues of a living death or a dead life" [diese wahren Standbilder eines lebendigen Todes oder eines toten Lebens].[12]

The perception of the automaton as "living death or a dead life" is precisely what Masahiro Mori posited in "The Uncanny Valley" (1970) as the extreme of revulsion, repugnance, or fear. As robots become more humanlike, Mori hypothesized, they will appear progressively more familiar until the encroachment on human identity renders them threateningly *unheimlich*. The deepest nadir of that "uncanny valley" of rejection occurs when an android seems to be possessed by a haunting spirit or a zombie-like creature animated by human-like drives and impulses. Had he read Hoffmann or at least Freud's reading of Hoffmann, Mori would have approved Hoffmann's suggestion that young ladies cough at regular intervals so that they are not mistaken for automata.[13] The problem occurs when the perception of the non-human disrupts the verisimilitude, and can be resolved when a more harmonious verisimilitude is achieved.

Mori's graph of the "uncanny valley" indicated a gradual rise of attraction, a steep descent into revulsion, followed by a correspondingly near vertical rise to harmonic attraction. The doll designed for sexual functions required only the anatomical features, not the artificial intelligence. After Nathanael's nightmare in which Coppelius steals Clara's eyes, he sees only death peering back at him from beneath her eyelids; and when she tells him he ought to burn the poem relating this nightmarish tale, he rejects her as a "lifeless damned automaton" [Du lebloses, verdammtes Automat!].[14] Equipped with one of Coppola's small telescopes, Nathanael becomes entranced in spying on Olimpia. He initially responds to her fixed and unseeing eyes as utterly "uncanny" [unheimlich].[15] Further attention, however, reverses the impression. His gaze animates her gaze:

> Now Nathaniel first gazed upon Olympia's beautifully formed face. Only the eyes seemed to him peculiarly fixed and dead. Yet as he peered more and more keenly through his telescope, it seemed as if moist moonbeams began to shine in her eyes, as if the power of sight was awakened. Ever more lively blazed the looks. Nathaniel lay at the window as in a magical trance, gazing on and on at the heavenly beautiful Olympia.

> [Nun erschaute Nathanael erst Olimpias wunderschön geformtes Gesicht. Nur die Augen schienen ihm gar seltsam starr und tot. Doch wie er immer schärfer und schärfer durch das Glas hinschaute, war es, als gingen in Olimpias Augen feuchte Mondesstrahlen auf. Es schien, als wenn nun erst die Sehkraft entzündet würde; immer lebendiger und lebendiger flammten die Blicke. Nathanael lag wie festgezaubert im Fenster, immer fort und fort die himmlisch-schöne Olimpia betrachtend.][16]

When Ludwig described his reaction to the figures in the wax museum, Hoffmann had him cite the words of Macbeth when confronted by the ghost of Banquo: "Thou hast no speculation in those eyes / Which thou dost glare with" (3.4.95–96). Hoffmann cited the passage again (e.g., in "Automata,"[17] and *Life and Opinions of the Tomcat Murr*[18]), and even more frequently he repeated the haunting image of the sightless eyes. A challenge for the maker of the automaton is to create an illusion of sentience in the glass or porcelain orbs planted in the carved or molded head. In "The Sandman" Nathanael's deranged imagination converts human to automaton and automaton to human. In "Automata" the mechanical prophet whispers words concerning Ferdinand's intimate secret. Does the explanation lie in a hoax perpetrated by Professor X? Some natural or occult phenomena? A psychological projection such as triggered Nathanael's madness?

If it were a hoax, it was a good one, but not necessarily superior to Professor X's other clock-work gadgetry. In *Letters on Natural Magic* (1838), Sir David Brewster devoted an entire chapter to auditory illusions, including the speaking head, the invisible girl, and ventriloquist trickery,[19] and another chapter to the musical, dancing, writing, speaking automata of Vaucanson, Kempelen, Kratzenstein, and others.[20] Two of the auditory illusions are close to Hoffmann's mystery of the speaking Turk. One was the hoax contrived by Thomas Irson and exhibited in the court of Charles II: "The questions had been proposed to the wooden figure by whispering into its ear, and this learned personage had answered them all with great ability, by speaking through a pipe in the same language in which the questions were proposed."[21] The second was Kempelen's automaton: "The figure of a Turk was placed on a box, which was filled . . . with pipes, bellows, &c.; but the voice was communicated from a person in an adjoining room."[22] Philip Thicknesse[23] and Joseph Friedrich Freiherr von Racknitz[24] were among the critics who exposed Kempelen's trickery and showed that the wind-up mechanism was little more than a distraction. Hoffmann allows his characters Ludwig and Ferdinand to snoop and pry as sceptics, but neither exposes the hoax. No person is hidden within the box because there was no space for even the smallest of dwarves;[25] nor could the Professor have wrought the deception as ventriloquist, because he was usually engaged in conversation when the voice was heard.[26] Further, the voice displayed erudition and a command of several languages.[27] As with Kempelen's Turk, conversation was conducted by whispering into the ear.[28]

The whispered messages were uncanny because the automaton knows the secret of Ferdinand's love for a mysterious singer, a secret not even revealed to his friend Ludwig. Ferdinand asked the oracular automaton whether he would ever again experience a moment like the one in which he was happiest [Werde ich künftig noch einen Moment erleben, der dem gleicht, wo ich am glücklichsten war?].[29] The automaton tells him that he must first turn around the miniature portrait that hangs from his neck, which is now faced inward. Ferdinand obeys, and the Turk informs him: "When you see her again, you will have lost her" [Wenn du sie wieder siehst, hast du sie verloren].[30] Ferdinand himself drew the portrait. No one else had seen it. No one else knew that Ferdinand had listened enraptured as she accompanied herself on the fortepiano playing Franz Schubert's melody and singing the lyrics by Pietro Metastasio: "Remember, my beloved, if it happens that I die, / how much this faithful spirit loved thee" [Mio ben ricordati/s'avvien ch'io mora, / quanto quest' anima / fedel t'amò].[31] Ferdinand hears this melody again in the background as the Turk utters the fatal pronouncement.

Ludwig tries to find a rational explanation: perhaps the oracular words triggered the memory of the melody resident in his mind or in nature itself. Mind, nature, music, and machine are the four compelling sources of interconnected being. Music is the language shared by the other three. The machine is the metronome or heartbeat that provides the vibrating rhythm and pulse, mind the structure and composition, nature the melodies and harmonies. Ludwig's example is the Aeolian harp, the instrument that prompted Samuel Taylor Coleridge's speculation whether

> all of animated nature
> Be but organic Harps diversely framed,
> That tremble into thought, as o'er them sweeps
> Plastic and vast, one intellectual breeze,
> At once the Soul of each, and God of all? ("Eolian Harp,"
> lines 36–40)[32]

The Aeolian harps of nature are heard when the winds play upon the leafy branches of trees, upon the openings of cliffside caves, or upon man-made instruments with strings tuned to respond to the breeze in diatonic chords. The music which dwells within us, Ludwig speculates, may be the same as the music that lies hidden in nature. The spell may be lifted in dreams and "in the pure psychic efforts of the mind" [im reinpsychischen Wirken des Geistes].[33] Even in the sounds of familiar instruments in concert we hear the sounds of nature ["wir hören selbst im Konzert bekannter Instrumente jene Naturlaute"].[34] Ludwig goes on to declare that the Aeolian harp may be rightly praised for giving voice to the sounds of nature ["der Natur Töne"],[35] but in truth it is only a child's toy. The grander conception, described in the *Hamburgischer Correspondent* (1786, No. 160) and the *Göttinger Taschenkalender* (1789, 129–134), was the Meteorological Harmonica capable of responding to the subtler alterations of atmospheric pressure as well as greater powers of a storm. Fifty meters of "thick wires and cables are stretched across wide expanses in nature, and when they are set in vibration by the winds they resound with mighty chords" [Dicke in beträchtlicher Weite im Freien ausgespannte Drähte wurden von der Luft in Vibration gesetzt, und ertönten in mächtigem Klange].[36]

The Aeolian music of nature may be tranquil and even benevolent as the "intellectual breeze" that sweeps across "all of animated nature." But, as Coleridge also acknowledges, a "dull sobbing draft" also "moans and rakes / Upon the strings of this Æolian lute" ("Dejection: An Ode," lines 6–7). Just as the Meteorological Harmonica howls in torment under the storm, the simple lute, too, is capable of "a scream / Of agony by torture lengthened out," when played upon by the "Mad Lutanist" raving at the window ("Dejection: An Ode," lines 96–103). The scale from the gentle to the tempestuous melodies of nature is interpreted through the "psychic rapport" between two sensitive instruments: mind and harp. Ludwig suggests that the automaton, too, mediates the subtler excitations of our nerves much like the strings of the harp:

> It is the psychic power, which strikes the strings within us which otherwise hum in disharmony, and are thus made to vibrate and resound, so that we perceive the pure accord; we ourselves give answer through an inner voice that a foreign intellectual principle has aroused in us.
>
> [Es ist die psychische Macht, die die Saiten in unserm Innern, welche sonst nur durcheinander rauschten, anschlägt, daß sie vibrieren und ertönen und wir den reinen Akkord deutlich vernehmen; so sind wir aber es selbst, die wir uns die Antworten erteilen, indem wir die innere Stimme durch ein fremdes geistiges Prinzip geweckt.][37]

Citing the opening chapter of G. H. Schubert's *Views from the Nightside of Natural Science* [*Ansichten von der Nachtseite der Naturwissenschaft*, 1808],[38] Ludwig repeats the theory that early mankind lived in harmony with nature, attuned to its music, and responded with a deep appreciation of the music that came from without and within.[39]

In describing how the mind may establish a rapport with nature, musical instruments, and machines, Hoffmann has Ludwig adopt from animal magnetism references similar to those recurring in his tale "The Magnetist" ["Der Magnetiseur," 1815]:[40] "psychic influence" [psychischen Einfluss], "pure psychic-effective natural energies" [rein psychisch wirkende Naturkräfte], and "the mysterious depths of psychic effects" [die geheimnisvollen Tiefen der psychischen Einwirkungen].[41] In that tale, too, Hoffmann contrasted the treatment by gentle transition vs. hysterical crisis, as represented respectively in his tale by Theobald and Alban. For his repeated reference of "psychic rapport" in "Automata," Hoffmann may have relied on Carl Adolph Eschenmayer's popular work (1816) on the apparent magic of animal magnetism in its psychic and physiological effects. Rapport, as Eschenmayer defined it, was "a mental mating of the organ of the soul with the senses" [eine geistige Begattung des Seelenorgans und des Gefühlvermögens].[42]

Poetry, art, music, and song developed in humanity's self-celebration. Inspired by nature, the creative endeavor soon turned to rivalling nature. Acknowledged as a masterpiece of mechanics and acoustics,[43] the Turk of Professor X was presented as significantly more: a creature capable of sentience, clairvoyance, prophecy, and "psychic rapport." As the appearance and capabilities of an automaton are made more human, Masahiro Mori argued, an observer's emotional response will become increasingly positive and empathic. Inevitably in that progress a point is reached in which its mechanical identity is betrayed by its human appearance, looks and actions—or vice versa. At this point the lifelike illusion is perceived as a deception—a potentially dangerous deception. The turning point in "Automata" occurs when the Turk foretells Ferdinand's fate. In "The Sandman" it occurs when Nathanael's beloved Olimpia is wrenched apart by Spalanzani and Coppola. The difference is that in the latter tale, Hoffmann has shared with readers all the details of Olimpia's mechanical identity; in the former, Hoffmann has concealed from readers the source of the Turk's voice and knowledge.

The human shape given to Olimpia temporarily deceives Nathanael until her mechanical workings are violently exposed. Worse than being deceived, Ferdinand and Ludwig are alarmed by the seemingly oracular revelations of the Turk. A strong revulsion, Mori argues, is only a negative interim in the response to robotic development. Let further advances be made, so that the robot's facial expressions and body movements become less distinguishable from that of a human being, then the emotional response becomes positive once again and approaches normal levels of social empathy. On Mori's graph, the level of positive response to the humanoid robot during the first ascendant phase is regained in the rise out of the uncanny valley when the response in the second ascendant phase recognizes the animated figure as a Bunraku puppet.

Inspired by Japanese Bunraku, the puppetry in the Disney musical, *The Lion King* (1999), overcame the uncanny effect because the puppeteer was a visible part of the puppet. The hornbills, the eighteen-foot giraffes, the hyenas, the stampede of wildebeest, all these puppet animals were intriguing because the audience beheld the doubleness of the characters as animal and human at the same time. To adapt William Butler Yeats's terpsichorean riddle to a kindred artistic mode and medium: "How can we know the puppeteer from the puppet?" Once we recognize the android-maker as a persistent part of the android's functions, the uncanniness subsides. In *The Wizard of Oz* (MGM, 1939) the frightening intimidations of the giant head are eradicated in the moment that Toto pulls back the curtain and Frank Morgan, in his role as wizard, tries to restore authority with his ineffectual command, "Pay no attention to the man behind the curtain." In overcoming the uncanny, the "man behind the curtain" must be moved front and center.

This is precisely what Hiroshi Ishiguro[44] accomplished with the project he introduced in June 2006 at the ATR Intelligent Robotics and Communication Laboratories in Japan. His assistant pulled back a curtain to reveal a *Doppelgänger*, a second Hiroshi Ishiguro dressed identically, but identifying itself as Geminoid HI-1. Emphasizing his role as "man behind the curtain," Hiroshi Ishiguro has offered interactive exhibitions enabling participants to engage in dialogue with the robots. Witnessing the mechanics and artistry in animating the lifelike replica provides a familiarity that dispels the creepiness, as if the ventriloquist has taught the dummy all of his art so that it may perform on its own.

Are we to believe then that the dance, a living being and the automaton in harmony together, would contribute to the rise out of the uncanny? Wouldn't the dance achieve precisely the harmonic interaction of the Bunraku puppets that Mori stipulated? It might, but only with awareness. Akin to distinguishing friend from foe, the ability to detect the disparity between person and machine is a natural instinct. Nathanael in his derangement lost that protective sense. Having considered Olimpia's seemingly sightless eyes uncanny, Nathanael went on to imagine that they gained a glimmer when he stared into them longer.

Spalanzani holds a debutante ball at which Olimpia plays the harpsichord, sings and dances. Nathanael is enraptured and requests a dance. Taking her ice-cold hand, he feels himself trembling from a horrible death-frost [von grausigem Todesfrost],[45] but as soon as he sees the reflection of love and longing in her eyes, he feels her hand warming to his touch and a pulse beating to his own:

> He embraced the beautiful Olympia and flew with her through the rows.—He thought he had always danced in rhythmic time, but when

> his step was often disrupted by the rhythmic precision of Olympia's dance, he soon noticed how much he lacked the beat.
>
> [er umschlang die schöne Olimpia und durchflog mit ihr die Reihen.—Er glaubte sonst recht taktmäßig getanzt zu haben, aber an der ganz eignen rhythmischen Festigkeit, womit Olimpia tanzte und die ihn oft ordentlich aus der Haltung brachte, merkte er bald, wie sehr ihm der Takt gemangelt].[46]

Nathanael is so preoccupied with Olimpia that he fails to notice the guests suppressing their laughter as they watch the strange dance.

In "Automata" Hoffmann has Ludwig register a far harsher judgment. For him the sight of the mismatched dancers, even without the lover's passion, would arouse not amusement but disgust:

> The very connection of a person with a dead figure, one that merely apes the human form and motion with the same deeds and drives, has for me something depressing, uncanny, dreadful. I can imagine that it must be possible, by means of gears hidden inside, to let figures dance nimbly and artfully, these figures must also be able to perform a dance together, and to turn and twist in all kinds of circles, so that the living dancer might take hold of the dead wooden female dancer and whirl around with her. Would you be able to bear such a sight for a single minute without inner horror?
>
> [Schon die Verbindung des Menschen mit toten das Menschliche in Bildung und Bewegung nachäffenden Figuren zu gleichem Tun und Treiben hat für mich etwas drückendes, unheimliches, ja entsetzliches. Ich kann mir es denken, daß es möglich sein müßte, Figuren vermöge eines im Innern verborgenen Getriebes gar künstlich und behende tanzen zu lassen, auch müßten diese mit Menschen gemeinschaftlich einen Tanz aufführen und sich in allerlei Touren wenden und drehen, so daß der lebendige Tänzer die tote hölzerne Tänzerin faßte und sich mit ihr schwenkte, würdest du den Anblick ohne inneres Grauen eine Minute lang ertragen?][47]

Ludwig expresses concerns about the unnatural horrors of an automaton that so successfully counterfeits the appearance and movements of an actual woman that a living man might be seduced into taking the lifeless replica into his arms. None of the elements of Bunraku puppetry enter into Ludwig's judgment. Instead of artistry, ingenuity, and craftsmanship, he imagines only substituting a real partner for an artificial one.

That substitution of artifice for reality is further enhanced by Nathanael's overactive projection of his own warmth and passion into the wind-up doll as the object of his passion. Nathanael has made himself thoroughly aroused:

> He kissed Olympia's hand, he leaned to her mouth: ice-cold were the lips that met his glowing desire.—Just as happened when he touched Olympia's cold hand, he felt a momentary terror and thought of the tale of the dead bride, but Olympia had gripped him tightly and in their kiss her lips seemed to warm with life.
>
> [er küßte Olimpias Hand, er neigte sich zu ihrem Munde, eiskalte Lippen begegneten seinen glühenden!—So wie, als er Olimpias kalte Hand berührte, fühlte er sich von innerem Grausen erfaßt, die Legende von der toten Braut ging ihm plötzlich durch den Sinn; aber fest hatte ihn Olimpia an sich gedrückt, und in dem Kuß schienen die Lippen zum Leben zu erwarmen.][48]

In alluding to Goethe's ballad, "Die Braut von Korinth" (1797), Hoffmann prompts his readers to consider the coupling between the vampire and her victim as parallel to Olimpia and Nathaniel. Echoing Ludwig's rejection of the living person and lifeless automaton as an uncanny bond, Nathaniel's friend Siegmund cautions him that she is in fact uncanny in all of her manners and movement:

> Yet it is strange that many of us have reached the same opinion about Olympia. She seems to us—don't take it badly, brother!—in an odd manner stiff and without a soul. It is true that her stature and her face are regular. She could be judged beautiful, if her look wasn't so lacking any spark of life that she seems without the power to see. Her step is peculiarly measured. Every movement seems to be the result of wound up mechanical spring. Her playing, her singing have displeasing, dispassionate regularity as if she were a singing machine, and the same is with her dance. For us this Olympia has become thoroughly uncanny, and we don't want anything to do with her. For us it is as if she merely acts like a living creature, and yet has a story of her own.
>
> [Wunderlich ist es doch, daß viele von uns über Olimpia ziemlich gleich urteilen. Sie ist uns—nimm es nicht übel, Bruder!—auf seltsame Weise starr und seelenlos erschienen. Ihr Wuchs ist regelmäßig, sowie ihr Gesicht, das ist wahr!—Sie könnte für schön gelten, wenn ihr Blick nicht so ganz ohne Lebensstrahl, ich möchte sagen, ohne Sehkraft wäre. Ihr Schritt ist sonderbar abgemessen, jede Bewegung scheint durch den Gang eines aufgezogenen Räderwerks bedingt. Ihr Spiel, ihr Singen hat den unangenehm richtigen geistlosen Takt der singenden Maschine und ebenso ist ihr Tanz. Uns ist diese Olimpia ganz unheimlich geworden, wir mochten nichts mit ihr zu schaffen haben, es war uns als tue sie nur so wie ein lebendiges Wesen und doch habe es mit ihr eine eigne Bewandtnis.][49]

Nathanael is not deterred by "cold prosaic" friends who consider Olimpia uncanny. Her supposedly lifeless eyes have looked upon him with love radiating sense and thought. The poetry that Clara had rejected, Olimpia attended with patience and has favored with her wonderous "Ach! Ach!" She has few words, Nathanael admits, but these few are "true hieroglyphs of her inner world, full of love and higher knowledge" [echte Hieroglyphe der innern Welt voll Liebe und hoher Erkenntnis].[50]

In neither of these two tales does Hoffmann consider the creative bonding of the maker with the automaton to be a way-station in the ascent out of the uncanny valley. For that to happen, the exposition of the automata would have to be as open and interactive as the Bunraku puppeteers with their puppets. As a showman who derives his success from illusion and trickery, Professor X must remain "the man behind the curtain." By happenstance he witnessed the inception of Ferdinand's love for his ward at the inn and is able to exploit Ferdinand's secret by having it revealed by his oracular Turk. He has no intention of altering the marriage arranged for his ward. Unlike Professor X, Spalanzani has no interest in tricking the crowd. He is sufficiently satisfied in showing off his clock-work ingenuity, and he sees his reputation enhanced by Nathanael's thorough infatuation with Olimpia as if she were a living being.

When Ludwig and Ferdinand are invited by Professor X to view his private cabinet of automata, they expect to be initiated into the secrets of their workings, like the apprentice at Sais in the tale by Novalis. Instead, they are left to curse the Professor for leaving them as much in the dark as before.[51] Ludwig, who resents dancing with the dolls, resents the underlying trickery even more, and he contrasts the concealed puppetry with the open interaction that he experienced as a child with his own nutcracker. Its special features were the protruding eyes which would roll when cracking a hard nut, giving the entire figure such droll liveliness that the child could play with it for hours. In his hands the wooden dwarf was transformed, like the mandrake root, into a lifelike being. In comparison to his marvelous nutcracker the most perfect marionettes seemed stiff and lifeless.[52] Ludwig described being taken as a child to wax museum. "I liked my nutcracker much better" [Mein Nußknacker war mir doch lieber],[53] he shouted as he fled from the uncanny figures and their sightless eyes.

In Hoffmann's "The Nutcracker and the Mouse King," Fritz registers much the same critique for much the same reason. As his annual gift on Christmas Eve, Godfather Droßelmeier displays for the children a grand castle with clock-work animation. Fritz complains to his sister Marie that it is only meant to be looked at before it is then taken away. For the simple reason that they can do with the gifts what they want, Fritz declares: "I like much better what Papa and Mama give us" [da ist mir denn doch das viel lieber, was uns Papa und Mama einbescheren].[54] In response to Fritz's requests that the figures do something other than repeating their actions over and over, Droßelmeier concedes that other movements are not possible: "once the mechanism is constructed, so it must remain" [wie die Mechanik nun einmal gemacht ist, muß sie bleiben].[55] Just as he wished, Fritz receives a cavalry squadron of Hussars, brightly decorated tin soldiers that he can command and parade as he wants. In addition to the beautiful dress that she desired, Marie is also given the care of the nutcracker necessary to the Christmas festivities, but Fritz gives it such heavy duty with the largest and hardest of the nuts that it soon loses a few of its teeth and its lower jaw seems out of joint. Marie bandages the wounded nutcracker, which comes alive by night and defeats the evil Mouse King in battle.

As Hoffmann describes play, the child creates a fantasy plot which unfolds spontaneously with the child's own participation in the action. Fritz's play with the tin soldiers is much like Ludwig's play with the nutcracker, but Marie with the nutcracker is more like Nathanael with Olimpia, for her nutcracker is brought to life. After suggesting that the transformation may occur only in Marie's mind, Hoffmann then describes her as transported to a fantasy realm, like Anselmus in "The Golden Pot" ["Der goldne Topf," 1815]. The dream fantasy overwhelms reality in the conclusion, for the nutcracker is transformed by Marie's love into a prince, who takes her as his princess to a magical kingdom of dolls.

Upon first seeing Professor X's Turk roll its eyes, Ludwig is reminded of his nutcracker. The principal difference, like that of the Bunraku puppet, is situated in the personal interaction. As a child he knew precisely how to move and manipulate his nutcracker to achieve a full repertory of effects. With all mechanical and acoustic workings hidden from spectators, the Turk's movements and utterances remain uncanny. As Ludwig stands amidst the audience gathered to see the mysterious Turk, he repeats his childhood critique: "I liked my nutcracker much better" [Mein Nußknacker war mir doch lieber].[56]

After Coppola struggles with Spalanzani and runs off with the broken automaton, Nathanael goes mad at the sight of Olimpia's eyeballs rolling on the floor. "The Sandman" ends with Nathanael's brief reconciliation with Clara, permanently destroyed when he again feels threatened by Coppelius and attempts to throw Clara from a high tower before leaping to his death. In "Automata" the mechanical Turk's prophecy comes true: Ferdinand again sees his love, just as her marriage ceremony ends. Not the automaton but its operator, Professor X as con artist, is the true source of the uncanny. In both tales Hoffmann provides a critique of early nineteenth-century society. Nathanael's fate, he suggests, might have been prevented if lovers would ensure that their partners were not automata.

Tracking the overlap of *heimlich/unheimlich* remains indispensable to the critical use of the terms. In appropriating "The Sandman" as his case study, Freud repeats Clara's diagnosis of Nathanael's projection onto others of his own fears and desires. Mori's graph of the Uncanny Valley was designed to chart the successes and setbacks in android robotics in the 1970s, but it is as applicable to Hoffmann's automata as it is to current science. The frustrations and torment for Ludwig and Ferdinand, no less than the self-delusion of Nathanael, may all be traced to the crucial element of awareness and interaction that Mori credited to Bunraku puppeteering. Hoffmann's comic satire was relevant in an age in which bourgeois manners were becoming increasingly formal and all aspects of industry and commerce more mechanized. At the beginning the previous century, Susanna Centlivre, in her comedy *Love at a Venture* (1706), could provoke her audiences to laughter over the pretension of a character named Wou'dbe, "A Silly, Projecting Coxcomb," whose projections included city transportation by means of moving streets set into motion by clockwork (1.1.248–262). Three centuries later moving sidewalks are no longer wild fictions. With clockwork replaced by computer technology, and artificial intelligence driving cars and operating assembly lines, society is closer than ever to encountering the human replica that might not be recognized as a replica. Certainly, the examples abound in film and television, in computer animation and gaming. The human as machine and the machine as human are already among us. More than ever we must pay attention to the man behind the curtain.

NOTES

1. E.T.A. Hoffmann, *Sämtliche Werke*, vol. 4, *Die Serapions-Brüder*, ed. Wulf Segebrecht w/ collab. Ursula Segebrecht (Frankfurt am Main: Deutscher Klassiker Verlag, 2001) [=*DKV* 4].
2. E.T.A. Hoffmann, *Sämtliche Werke,* vol. 3, *Nachtstücke, Klein Zaches, Prinzessin Brambilla: Werke 1816–1820*, ed. Hartmut Steinecke w/collab. Gerhard Allroggen (Frankfurt am Main: Deutscher Klassiker Verlag, 1985 [2009]) [=*DKV* 3].
3. Sigmund Freud, "Das Unheimliche," *Imago. Zeitschrift für Anwendung der Psychoanalyse auf die Geisteswissenschaften* 5 (1919): 297–324.
4. Masahiro Mori, "The Uncanny Valley," trans. Karl F. MacDorman and Takashi Minato, *Energy* 7.4 (1970): 33–35.
5. Freud, 303 (my emphasis).
6. With regard to Vaucanson's automata, please see the contribution to this volume by Wendy C. Nielsen.
7. Regarding Jaquet-Droz's trio of automata, especially the female harpsichordist, please see Erin M. Goss's contribution to this volume.
8. *DKV* 4, 418.
9. Gilbert Ryle, *The Concept of Mind* (Chicago: University of Chicago Press, 1949), 15–16.
10. Gabby Wood, *Living Dolls: A Magical History of the Quest for Mechanical Life* (London: Faber & Faber, 2003), 34.

11. Please see the contribution to this volume by Wendy C. Nielsen regarding the Chess-Playing Turk of Wolfgang von Kempelen.
12. *DKV* 4, 399.
13. *DKV* 3, 46–47.
14. *DKV* 3, 32.
15. *DKV* 3, 25.
16. *DKV* 3, 36.
17. *DKV* 4, 399.
18. Hoffmann's second novel, *Lebensansichten des Katers Murr*, 2 vols., 1820–1822, *DKV* 5, 420.
19. Sir David Brewster, *Letters on Natural Magic* (London: John Murray, 1838; 5th ed. 1842), 157–178.
20. See Brewster, 207–211, 264–296.
21. Brewster, 160.
22. Charles Wheatstone, *The Scientific Papers of Sir Charles Wheatstone* (London: Taylor and Francis, 1879), 351. See also Brewster, 211–225.
23. Philip Thicknesse, *The Speaking Figure and the Automaton Chess Player, Exposed and Detected* (London: Printed for John Stockdale, 1784).
24. Joseph Friedrich Freiherr von Racknitz, *Ueber den Schachspieler des Herrn von Kempelen, nebst einer Abbildung und Beschreibung seiner Sprachmachine* (Leipzig: J. G. Müller, 1784; 2nd ed. Leipzig: J. G. I. Breitkopf, 1789).
25. *DKV* 4, 397.
26. *DKV* 4, 398.
27. *DKV* 4, 398–399.
28. *DKV* 4, 402.
29. *DKV* 4, 408.
30. *DKV* 4, 408.
31. *DKV* 4, 405.
32. Samuel Taylor Coleridge, *Coleridge: Selected Poems*, ed. Richard Holmes (New York: HarperCollins, 1996).
33. *DKV* 4, 423.
34. *DKV* 4, 423.
35. *DKV* 4, 423.
36. *DKV* 4, 423–424.
37. *DKV* 4, 414.
38. Gotthilf Heinrich Schubert, *Ansichten von der Nachtseite der Naturwissenschaft* (Dresden: in der Arnoldischen Buchhandlung, 1808).
39. *DKV* 4, 421–422.
40. E. T. A. Hoffmann, *Sämtliche Werke*, vol. 2/1, *Fantasiestücke in Callot's Manier: Werke 1814*, ed. Hartmut Steinecke with collab. of Gerhard Allroggen and Wulf Segebrecht (Frankfurt am Main: Deutscher Klassiker Verlag, 1993) [=*DKV* 2/1, followed by page number].
41. *DKV* 2/1, 195–196.
42. Carl Adolph Eschenmayer, *Versuch die scheinbare Magie des thierischen Magnetismus aus physiologischen und psychischen Gesezen zu erklären* (Stuttgart und Tübingen: J. G. Cotta'schen Buchhandlung, 1816), 165.
43. *DKV* 4, 410.
44. Hiroshi Ishiguro, Hiroshi Ishiguro Laboratories, Osaka, 2016, accessed March 27, 2019, http://www.geminoid.jp/en/index.html and Miraikan National Museum of Emerging Science and Innovation, accessed March 27, 2019, http://www.miraikan.jst.go.jp/en/exhibition/future/robot/android.html.

45. *DKV* 3, 39.
46. *DKV* 3, 39
47. *DKV* 4, 418.
48. *DKV* 3, 40.
49. *DKV* 3, 41–42.
50. *DKV* 3, 42.
51. *DKV* 4, 418.
52. *DKV* 4, 409.
53. *DKV* 4, 410.
54. *DKV* 4, 243.
55. *DKV* 4, 246.
56. *DKV* 4, 410.

2

THE (RE-)WINDING OF HOFFMANN'S AUTOMATA

From Offenbach's 1881 Opera to Powell and Pressburger's 1951 Film

Ashley Shams

"IT'S AUTOMATIC! IT'S AUTOMATIC!" sings Hoffmann in horror at the sight of his beloved and now dismembered Olympia at the end of Act I in Powell and Pressburger's 1951 film adaptation of Offenbach's opera, *Les contes d'Hoffmann*. The music has stopped and the only thing audible is the sound of the automaton's eyes blinking, a mechanical sound much like the flapping of a film strip in a projector at the end of a movie.[1] There seems to be a connection. As Babington and Evans have stated, the English translation (above) of the French "c'est un automate" is more ambiguous and can be seen to apply to both the doll and the filmic apparatus.[2] This chapter takes its cue from their observation and argues that the film apparatus and the automaton share in their mimetic and automatic nature. As part of this study I will also demonstrate that the shift in genre that occurs in the adaption, from stage opera to opera film, captures the essence of automata in a more immediate way and strengthens Olympia's status as automaton.

The history of Jacques Offenbach's opera *Les contes d'Hoffmann* is a complicated one, assembled and disassembled much like the automaton itself. It is an adaptation of the fantastical drama, *Les contes d'Hoffmann*, written by Jules Barbier and Michel Carré for the Théâtre de l'Odéon in 1851. This play draws on several of E.T.A. Hoffmann's tales but, using a frame story, inserts the character of Hoffmann himself as the narrator/protagonist. The play incorporates mainly the following five Hoffmann stories (roughly in the order in which their events appear in the film): "Die Gesellschaft im Keller," "Klein Zaches, genannt Zinnober," "Der Sandmann," "Das verlorene Spiegelbild," and "Rath Krespel."[3] The play tells the tale of the protagonist's failed loves, three of which are nested in the stories (Olympia, Giulietta and Antonia) and one (Stella) occurs in the framing narrative. Heather Hadlock's meticulous research into the background of the opera reveals that it was the character of Dr. Miracle, written for the play, that became most memorable and that this influenced depictions of the opera at the time of its opening. The Doctor and Antonia appear second only to the image of Hoffmann in the 1881 engraving for *Le Théâtre illustré*. Olympia, sketched lightly, is featured in the upper right- and left-hand corners of the image.[4] By the 1951 release of Pressburger and Powell's film, however, it is Olympia who is featured most prominently on the official film poster, a shift that suggests the film's more pronounced connection to automata that will be discussed later in this essay.

Offenbach died in October 1880, leaving the opera seemingly unfinished just a few months before its scheduled premiere in February 1881. Léon Carvalho, the director of the Opéra-Comique, and the composer, Ernest Guiraud, took over. Rehearsals continued, but just a few days before opening night it was deemed to be too long and Carvalho made the decision to cut the entire fourth act, the tale of Giulietta. With this decision, several pages were literally cut from the manuscript while other sections were reordered and dispersed throughout the remaining acts. The cut pages were placed in a box and eventually lost, leading to much of the later confusion about authorship and the composer's original intent. Within a year of the modified premiere, the Giulietta act was restored. Giraud, however, chose to place it in between the Olympia and Antonia acts, instead of Offenbach's intention of positioning it as the last of the three, and it is this order that became the standard for the next seventy-five years. The next substantial change was orchestrated by Raoul Gunsbourg, director of the Monte Carlo Opera, who decided to recompose the Giulietta act and return it to prominence for the 1904 production. The changes he made included the addition of the Hoffmann inspired "Scintille Diamant" and Gunsbourg's own music, as in the case of the famous septet "Hélas, mon coeur s'égare encore." Referring to Debussy's 1904 review of the Monte Carlo production, arts journalist Donatienne Graham-Bates paraphrases his words with, "Of course, Offenbach is dead. M. Gunsbourg fancies himself as Offenbach. Luckily Offenbach is dead. If not how could Offenbach have written such a score without M. Gunsbourg."[5] It seems as if, from its troubled beginning, the staging of the opera, in addition to the Hoffmann-inspired content, lent itself to a certain doubling and feeling of the uncanny. Offenbach, the father of the opera, has "returned from the dead" in the form of his musical double Gunsbourg, who, while not Hoffmann, has replaced him. Hadlock furthers this notion of doubling and authorship when she points out that from its opening in 1881 to the 1907 publication of the Gunsbourg edition, *Les contes d'Hoffmann* gained—or regained—an entire act, a palimpsest of music by Offenbach, music in imitation of Offenbach, and one number, the septet, that had no connection at all with Offenbach."[6]

These palimpsestic qualities reverberate throughout the 1951 film adaptation, *The Tales of Hoffmann* by Michael Powell and Emeric Pressburger. By its very nature adaptation is built on a prior work. Linda Hutcheon characterizes adaptations as an extended intertextual engagement with the adapted text. They require an acknowledged transposition of a recognizable other work or works, which is also a creative and interpretive act of appropriation. She concludes that "an adaptation is a derivation that is not derivative—a work that is second with being secondary. It is its own palimpsestic thing."[7]

Sir Thomas Beecham, of the Royal Philharmonic at that time, proposed the idea of the film adaption to Powell and Pressburger.[8] Beecham wanted to make opera, which was dominated by German music in 1950s England, more accessible and appealing to the general public. He had collaborated with the filmmakers earlier on a very commercially successful film, *The Red Shoes*. Beecham arranged the Offenbach opera for the new film and added the dragonfly ballet that appears in the prologue. As with adaptations, things were familiar yet different. The libretto was sung in English but the original notes rang through. Most of the cast had appeared together in *The Red Shoes* in very different roles. Other changes that occurred in the adaptation and that bore relevance to the Olympia act included the recasting of Olympia as a ballet-dancing *and* opera-singing doll, the use of the prerecorded soundtrack, and the addition of a final scene that occurred outside the story.

As Hutcheon points out, "most commonly considered adaptations are those that move from the telling to the showing mode, usually from print to performance."[9] In the case of Offenbach's opera this is true, albeit one step removed, since we would first have to return to Barbier and Carré's play in order to arrive at the first adaptation of the source material, Hoffmann's short stories. Both the play and the opera are said to be in the showing mode, but showing in film is different than showing in theatre. Film by its very nature favors the recording of movement. Siegfried Kracauer identifies three types of movement that the recording function of film privileges: the chase, dancing, and nascent movement. The translation of Olympia into a dancing automaton would seem to fall into the second category; however, the dancing to which Kracauer refers is one of natural movement that emerges realistically from everyday life and the physical reality of the characters' lives, more in the style of impromptu performances than stage choreography. He disparages "canned" recordings of dance, explaining that "screen reproductions of theatrical dancing either indulge in a completeness which is boring or offer a selection of attractive details which confuse in that they dismember rather than preserve the original."[10] While Olympia's act features ballet, it is not the equivalent of a stage performance. It escapes this categorization because Olympia's dancing is part and parcel of her physical reality: she is dancing for Hoffmann and it is thus a natural part of this adaptation's story. Kracauer's third type of movement, nascent motion, is of particular importance here because it contrasts movement with that which is motionless.[11] It is in Olympia's halting by contrast to her movement that her identity as automaton is made manifest. For Kracauer, when characters return to motion, they take on life again. Olympia is an inanimate object when still but alive when dancing. When Olympia's winding mechanism runs out and she stops moving, her body is shown in a bent or doubled-over position, like a lifeless doll.

Film's emphasis on movement and the visible necessitates a bodily equivalent to the voiced coloratura. Frederick Ashton's choreography achieves this while furthering Olympia's identity as automaton through the use of pointe work, arabesque, and pirouette. All three seem to defy gravity and therefore go beyond "human" boundaries. The rising voice is usually accompanied by rises on pointe and arms above the head in fifth or fourth position. The arabesque is a position of balance, strength, and flexibility and it requires the dancer to balance on her standing leg while raising her free leg, kept straight and turned out, to a ninety-degree angle behind her. Olympia's pirouettes, sometimes more than six in a row on pointe, are virtuosic and reminiscent of the dancing dolls in music boxes.

In watching Olympia's act, even casual fans of ballet cannot help but detect elements of two key earlier ballets, Arthur Saint-León and Charles-Louis-Étienne Nuitter's *Coppélia* (1870) and Michel Fokine's *Pétrouchka* (1911). The connection to the former is clear: *Coppélia* was inspired by two of Hoffmann's stories, "Der Sandmann" and "Die Automate," drawing particularly on the figure of a dancing automaton. The latter has commonalities in its use of dancing dolls and marionettes that come to life. In recognizing these two ballets that feature automata, more knowledgeable dance scholars might think of still others like *Arlequinada* (1900) and *Die Puppenfee* (1903), as does Linda Austin in her investigation of the popularity of this theme in ballets of the late nineteenth and early twentieth centuries. She identifies the emergence and proliferation of automata ballets with the scientific debate over human automatism that was investigating the human-machine distinction by questioning "whether life, specifically human life, was based on neuromotor activity or hinged on conscious cerebral involvement."[12] In her essay, Austin highlights the technological advancements showcased in Jacques de Vaucanson's flute player of 1738 and his defecating duck of 1739, because these mark the moment when machines began to simulate human organisms in their physical and physiological movement. She then juxtaposes this technological development with the ballet technique of the time, which rested on the converse notion that "through technique the human body could simulate the speed and exactness of a machine and simultaneously convey the spontaneity of a living organism."[13] In the context of the Olympia ballet, it is the focus on the machine-like movement that applies. The ballet *Coppélia*, on the other hand, contrasts the doll-like movements of Coppélia with movements that show signs of a living being in Swanhilda pretending to be Coppélia as she comes to life (hence the need for dance to convey the spontaneity of a living organism).

The scientific and philosophical debate over the genesis of movement was taken up in the dance world by Noverre and Blasis, both ballet maîtres, who argued that the mind or soul was the catalyst of movement. For Noverre, a dancer's movements that were not initiated by the intellect were to be equated to the mindless clumsy movements of marionettes (though he stressed, however, that the dancer's movements should appear void of any trace of this volition or effort). The "Olympia" act takes advantage of this distinction when it uses the clunky movements of marionettes (albeit "embodied" by human dancers) in contrast to the airiness of Olympia's untethered, seemingly spontaneous movements, reinforcing her double identity as animate/inanimate and as soulful as opposed to soulless machine. The surreal and unnatural make-up of the marionette dancers also serves to mark their status as artificial and rudimentary as opposed to Olympia's more natural and evolved look.

In the first half of the eighteenth century, the dancer in motion was described as "a complicated piece of machinery with component parts functioning independently of any central control" and the dancer's body, in the words of French physician and philosopher Julien Offray de La Mettrie, as "a self-winding machine."[14] It is no coincidence, then, that a few decades later the figure of the automaton would work itself into the ballet repertoire. But why would it appear again in the 1951 *The Tales of Hoffmann* and why is it worthy of discussion anew?

The figure of the automaton offers a vehicle for a meta-commentary on the film process as a whole. Both the film and Olympia share their mimetic nature. They mimic "reality" by deceiving our senses. The cinematic experience is perceptual, yet what we see is not what we perceive. In cinema, "the activity of perception which it involves is real (cinema is not a phantasy), but the perceived is not really the object, it is its shade, its phantom, its double, its *replica* in a new kind of mirror."[15] The image of Olympia on the screen is only a projection of a two-dimensional image of Moira Shearer playing the role of an automaton. The living and breathing Moira Shearer is not in front of us, although an inanimate replica of her is. Furthermore, projection itself is an automated process. The moving image that we see is powered by a projector, a mechanical machine with manmade moving parts. Both Olympia and the film have a captivated and believing audience as long as they are in good working order, but as soon as a mechanical breakdown occurs, the illusion is broken. "More than the other arts, or in a more unique way, the cinema involves us in the imaginary: it drums up all perception, but to switch it immediately over into its own absence, which is nonetheless the only signifier present."[16] This is equally true of Olympia. It seems she has beguiled Hoffmann in a fantasy of dance and song but her ability to sing and dance is what marks her as an automaton, a life-like machine, not a living human being. As an automaton, Olympia is signified precisely by her absence of life.

What is the effect of the film's emphasis on sound? Scholars such as Marcia Citron,[17] Kara Reilly,[18] and Lesley Stern[19] have addressed the separation of voice and sound created by the use of the prerecorded soundtrack in the film. Some of the effects addressed include the doubling of the onscreen actor/dancers with the off-screen singers, the anxiety caused by the disembodied female voice, the problematic lip-synching that disrupts the suspension of disbelief, and the rendering of onscreen performers into automata as they, in a sense, become robotic, responding and reacting to a voice that is outside of themselves and not their own. Moreover, Olympia's voice is further problematized when examined from a posthumanist perspective. Olympia sings an aria for Hoffmann. Singing, and the aria in particular, is often associated with self-expression but, as an automaton, Olympia has no real "self" to express. Indeed, in the original the words she sings refer to herself in the third person:

> Tout ce qui chante, résonne
> Et soupire tour à tour,
> Emeut *son* cœur qui frisonne
> D'amour! Tout parle d'amour!
> Ah! Voilà la chanson mignonne,
> La chanson d'Olympia!
>
> [Everything that in turn
> sings, resounds and sighs
> excites *her* heart, which quivers
> with love. That is the pretty song.
> Olympia's song!][20]

In the film version, the aria is sung in the first person but the question of voice as self-expression is further complicated by the fact that Moira Shearer's performance of the singing Olympia is not given in her own voice and thus is not the self-expression of the actress/dancer interpreting her role. Moira Shearer is not an opera singer, and even if she were, she still could not sing the aria while pirouetting! Her performance is augmented. The singing we hear is actually voiced by Dorothy Bond, off-stage and prerecorded. In a sense then, we are watching a hybrid performance made possible by a "prosthetic" voice.

Given all these twists and turns, why does sound and thus voice still play an important role and have such an emotionally charged effect on the audience? One reason for the importance accorded it, of course, stems from the fact that this is a film about opera. Added to that is the fact that the idea for the film came from the conductor and impresario Beecham himself. The *raison d'être* of the film originates in the musical world. One reason for its emotional impact may be the authenticity of sound itself. Sound is the only direct sensory perception involved in a film. Sound is reproduced, not copied. An image, on the other hand, is copied, transformed, and eventually decoded through our minds back into a representation of the original image. We never see the actual object itself. Jean-Louis Baudry likens the cinematic experience to Plato's cave.[21] According to this simile, the fire, located up and at a distance behind the prisoners' perspective, is the projector, the immobile prisoners are similar to the audience, and the shadow play they watch on the cave wall is like a movie projected onto the screen. He eventually arrives at Plato's treatment of sound. Plato resorts to sound as the final means to communicate to the prisoners the "reality" of what they are experiencing, and it is voice that offers the most convincing evidence because, as Baudry surmises, sound resists being caught up in simulacra. "[O]ne does not hear an image of the sounds but the sounds themselves."[22] Despite the almost hyper-visual stimulation of the film, it is its sound that is the most authentic and most immediate sensory experience. For this reason, the music and the singer's voice can provoke a strong emotional reaction in the listener.

The visual plane of the film offers other pleasures. The scopophilia of the source text is amplified in the *Tales*. Nathaniel[23] first falls in love with Olympia as he watches her through the spyglass he purchased from Coppelius. As Kara Reilly points out, he has "brought her to life with his obsessive gaze."[24] In the opera and the film, the spyglass is transformed into different spectacles that, once on, reveal to Hoffmann a world populated by inanimate objects come to life. This is particularly well communicated in the film because of the camera's ability to show character perspective, thus allowing us to see what Hoffmann sees. Reilly states that it is in this cinematic staging of Hoffmann's gaze that the film offers insights into scopophilia, the sexual pleasure derived by objectifying another through sight, and the eye as an erogenous zone.[25] We see his pleasure as he watches Olympia but we also watch Olympia through his eyes. As Laura Mulvey has shown, the film spectator, in sharing the male gaze of the protagonist, partakes in this same scopophilic pleasure of using another person as an object of sexual stimulation through sight.[26] Olympia becomes the object of our scopophilic gaze as well. Furthermore, just as Nathaniel and Hoffmann bring the inanimate Olympia to life through their gaze, we, the movie audience, also bring the *film* to life through ours. Without the viewer, a film is only a piece of celluloid or, in the case of current technology, a series of keystrokes producing a computer code. Without the spectator, a film is an inanimate object. The mechanized movement of the projector may show two-dimensional images made up of light and darkness but it is only through our watching of the film that the characters "come to life," so to speak.

The underlying psychology of the castration anxiety that Freud identified in "Der Sandmann" finds expression in the film through the medium's emphasis on viewing and, therefore, the eyes. It is also amplified through the problematic position of the female figure on display. While the woman on the screen may be the object of what Mulvey has termed "the male gaze," her lack of a penis implies "a threat of castration and hence unpleasure."[27] One of the ways to escape from the castration anxiety is through fetishistic scopophilia, which allows the represented figure to become a fetish in itself, thus making it more reassuring rather than dangerous. Mulvey links this to over-valuation and the cult of the female star. This is partly the case for Olympia because she is danced by Moira Shearer, whose star status was unquestionable by the early 1950s. Shearer had already starred in Powell and Pressburger's earlier "hit" film *The Red Shoes* which had been nominated for five Oscar awards in 1949, including best picture. As the first famous redhead ballet dancing movie star, she attained a cult status of her own, and she brought this renown with her to the *Tales*.

According to Mulvey, the other means to contain castration anxiety is through voyeurism. Voyeurism has associations with sadism since its pleasure lies in ascertaining guilt and asserting control over the guilty person through punishment or forgiveness. While Hoffmann enjoys watching Olympia on display, the sight of her also has the potential to evoke castration anxiety (through her lack of a penis). As a consequence of being the source of this anxiety, she must be punished. Olympia's violent dismemberment is the visual manifestation of this punishment. Hoffmann (and the viewer) watch as she is chased, caught, and torn apart by Coppelius. Through the use of cinematic trickery, we see her "dismembered" leg kicking, Cochenille (the puppet master) caressing her amputated hand, and finally a close-up of her "decapitated" head as springs pop out of it. The threat she represented is now contained and Hoffmann's power over her is ascertained. This punishment is doubly fitting since she has deceived Hoffmann from the outset. Her destruction absolves him from his earlier reproachful naiveté.

Reilly argues that Olympia is not just the perfect Freudian fetish object but a Marxist fetish as well.[28] In drawing a parallel between Olympia and Marx's example of the commodity fetish, she shows how his words depict a table almost coming to life, the inanimate becoming animate "through the psychic forces of desire and exchange." This she equates to Olympia who, also made of wood and brought to life through her creators' desire and social value, is subject to negotiation and exchange as they fight over who owns her. Commodification can also explain the popularization of automata in nineteenth century ballet. As Austin points out, it is likely that many of the audience members at *Coppélia*'s premiere owned the small-scale artisanal automata that were seen as luxury products. These "domesticated" automata were produced in large numbers for individual ownership. They were more like toys (for adults) than the original life-sized androids that had appeared at expositions earlier, which explains the audience's attraction to the ballet and the portrayal of Coppélia as more ludic than her mysterious and uncanny origins in Hoffmann.[29]

As an even more pressing example of a Marxist fetish object, twenty-four pages of Offenbach's original manuscript, including those pages that had been cut by Carvalho just before the 1881 premiere, were put up for auction by the French state in November 2002. Bidding started at 30,000 euros and the pages were purchased by musicologist Jean-Christophe Keck at 165,000 euros. During an interview just before the auction, Keck expressed concern that the manuscript would sell for an "outrageous price" due to auction house excitement where people may "pay anything to get it."[30] It is worth noting that the copyright had already been separated from the manuscript, so this was not the motivating factor behind the escalating bids. The sense that the value of the manuscript is no longer tangibly related to its production cost is furthered by the fact that a documentary film was made about the missing manuscript and its eventual sale.

The added final scene to the *Tales of Hoffmann* warrants discussion here. In this scene, the "film" has ended and the music has stopped. We see Sir Thomas Beecham as he puts down the conductor's baton and closes the score. In the next shot, we see a hand, presumably Beecham's, stamping "Made in England" on the back of the closed score or, in a sense, the final product—a final product derived from a French Romantic opera based on a German author's earlier stories. The spectator is left to wonder if this is a warning against the commodification of art in imitation of the satirical humor that comprises a significant part of Hoffmann's unique style.[31] Citron offers several interpretations, all of which lead her to conclude ironically, "The absurd juxtaposition tells us that art and commerce are not far apart, and that the film just witnessed is really great fun after all."[32]

But what is the source of this "great fun"? It originates in the film's use of Romantic irony. To understand its use here we must first examine its role in E.T.A. Hoffmann's original text. Maria Tatar argues that Romantic irony is present in both its overt and covert forms in "Der Sandmann." She explains: "a writer may use Romantic irony by destroying the fictional illusion through occasional authorial intrusion, or he may engage in what Friedrich Schlegel called a 'permanente Parekbase,' a perpetual destruction of illusion."[33] The first type, overt authorial interruptions, are the most ludic and in "Der Sandmann" include the narrator's direct address to the "dear reader" as well as his lamenting the difficulties he had in choosing an appropriate way to begin Nathanael's story. The second is more subtle and pervasive and "allows the author to break the illusion not by revealing himself in the text (in the form of authorial intrusion) but by concealing himself behind the text."[34] Tatar's analysis of the narrator's descriptions of Nathanael reveals that the latter's mental state mirrors his own to such a degree as to undermine our belief that they are two truly independent beings. The narrator is covertly reminding the reader that we are in the world of fiction and that he is the creator behind the character of Nathanael.

Instances of overt as well as covert Romantic irony occur in the film at both the narrative and filmic levels. Pressburger and Powell's *The Tales of Hoffmann* is a portmanteau film. The film, like the opera and play before it, uses a frame story and includes the character of Hoffmann himself as the narrator who tells the tales of his failed loves, which include the inner tales of Olympia, Giulietta and Antonia. Stella, his contemporaneous love interest, appears in the framing narrative. The film, however, omits the muse scene that traditionally concludes the epilogue. Each of the nested stories is introduced by a close-up of a storybook entitled *The Tales of Hoffmann* as its pages are being turned. The first page gives the name of the tale and the subsequent pages give the setting, the names of the characters and actors below photo stills from the act. As the "Tale of Olympia" begins, the audience learns that "the action takes place in Paris—before the Eiffel Tower was built." The tongue-in-cheek, nonfactual, fairy-tale-style "facts" are a reminder that we are about to enter into the world of fiction.

There is another possible explanation for the pleasurable, non-anxiety-provoking experience of the automaton by the spectator in this instance. If, as Frederick Burwick argues in the preceding chapter in this volume, Mori's trajectory of the automaton's uncanny effect is diminished when the puppeteer or android maker is made visible, then this would account for the ludic experience produced by the film.[35] Just as Burwick describes in his essay, it is through the doubling effect of both being present simultaneously that the spectator is intrigued rather than repulsed. In the case of the film, the doubling is produced by humans portraying marionettes and automata without much effort to conceal their underlying human nature.

Because this is a film, the function of narrator is also taken up by the filmic apparatus, and instances of authorial interruptions can occur at this level as well. Since, as discussed earlier, adaptations of print to performance move the narrative from the telling to the showing mode, the function of the narrator may be replaced by the features that allow a film to do the showing. The use of trompe l'oeil in the waltzing scene is one example. In this sequence Hoffmann and Olympia waltz down a magnificent and grand staircase with increasing frenzy. Cochenille, a puppeteer who is half-human, half-puppet, is waiting to catch them at the bottom. The scene is shot from above so that the film audience can experience the thrill of the trompe l'oeil staircase, which is really a carpet. The illusion of danger it presents thus provides another entertaining reminder that what is being shown is a fiction. Other examples include the transformation of Cochenille, played by Frederick Ashton, who in the next shot is "transformed" into an actual marionette. The continuity of his costume and make-up ensure that the viewer understands that the actor/dancer and the marionette are the same character. Hein Heckroth's set designs also contribute to the illusory undertone that permeates the film. The film features no "real" walls and instead relies on the use of transparent gauze, mobile backdrops, and abstract paintings, all of which help translate the textual elements of the fantastic to the film. Lesley Stern's description of the theatricality of the *mise-en-scène* can be applied to this discussion's use of covert authorial intrusion with its perpetual undermining of our belief that what we are witnessing is real. "The enchantment, and the terror too, derives from a region of instability, from the sometimes indiscernible difference between reality and illusion, stage and world, cinema and opera; but also between living and dead, animate and inanimate."[36]

The identity and position of the narrator in the film is multifarious. While both Nathaniel and the narrator provide the narrative within the literary text of "Der Sandmann," it is difficult to establish a parallel in "The Tale of Olympia" unless we acknowledge, as Tatar does for the original, that the protagonist and the narrator are one and the same. Nathaniel is a creation of the narrator's mind much like the character of the young student Hoffmann in "The Tale of Olympia" is a creation of the mature Hoffmann narrating from the outer frame story. The film reinforces the link between narrator and protagonist.

The final Beecham scene reveals yet another level of narration and offers an example of overt authorial intrusion that underscores the humorous elements that occur throughout the film. As with the original story, "[t]he appearance of a narrator reflecting on his creation constitutes the very essence of Romantic irony and serves as the matrix for comic elements embedded in the text."[37] The putting down of the conductor's baton is analogous to the writer who puts down her or his pen at the conclusion of a work. Here, as conductor (and it should be remembered that it was Beecham who first proposed the idea of making the film), he is in an authorial position at the extra-diegetic level and the film concludes with a Hoffmannesque desire to "instill in his readers the sense of joy he himself derived from understanding that a fictional text creates a world of illusion that must be kept apart from the reality of both author and reader."[38] Moreover, this final scene is akin to Burwick's earlier discussion (again in the previous chapter in this volume) of the uncanny being overcome as the "man behind the curtain" is finally revealed: it is Sir Thomas Beecham, whose musical orchestration "made" Olympia and the others dance and sing, who proves to be the ultimate behind-the-scenes "puppet master" of the film.

While the film ends on this playful note, it omits an important scene from the opera and in so doing leads to a pessimistic reading of the epilogue. In Offenbach's opera the muse appears in the epilogue and, like most nineteenth century muses, demands fidelity: "And I? I, the faithful friend whose hand dried your eyes? Thanks to whom pain, numbed, rises as dreams into the sky? Am I nothing? May the storm of passions be stilled in you! The man is no more, be reborn a poet! I love you, Hoffmann! Be mine!"[39] And, in the opera, Hoffmann accepts. But the muse is omitted in the film, thereby throwing into question the purpose of Hoffmann's suffering. If this is not the poet's sacrifice for his art, how are we to make sense of his suffering in the film? In many versions of the opera, Nicklausse, Hoffmann's confidante, reappears in the epilogue and reveals herself to the muse. An archival photograph from the film shows Pamela Brown, who played Nicklausse, painted in gold and wearing a muse's wreath and robes.[40] It is evident that Powell intended to keep the muse in the epilogue but that it was eventually edited out.[41] One is left to wonder if this artistic alteration was due to commercial demands, a hint of the commodification of art that Hoffmann decried.

Freud's discussion of "das Unheimliche" [the uncanny] opens another possibility for interpreting the final Beecham scene. In reexamining Freud's position on art, Adam Bresnick states that for Freud, art must be construed less as secondary representation of a given object or experience (mimesis) than as primary presentation of affect (poiesis). There is value in art through its cathartic potential. Bresnick summarizes art's catharsis as "the moment in which the subject's cathexis of the artwork reaches its apogee, only to subside as reason steps in to recuperate the aesthetic sublime as an object of retrospective critical cognition."[42] When a reader or viewer is completely absorbed in the story in front of her or him, an instance of the "uncanny" can occur when the reader/viewer is suddenly shaken out of that moment. "Intellectual uncertainty returns with a vengeance at this moment, for at stake is not merely uncertainty about a given plot, but a radical doubt about who is reading and what or whom is being read as the reader fantasmatically projects her or himself into the tale, dissolving the frame that would insure her or his ontological separation from the artwork."[43] In a sense the final Beecham scene produces just such an instance of the uncanny. The viewer has a sudden moment of uncertainty about the level (s)he actually occupies in the story as the shot of Sir Beecham pulls us into the outer frame. We for an instant are in his audience, which in turn means we have been part of the film.

But what would Hoffmann, a musician, composer and music critic in his own right, think of the film? Diana Stone Peters discusses Hoffmann's satire of the cultural philistine in her article "The Conciliatory Satirist." She states that, for Hoffmann, "[t]he man with pretensions to culture is not disparaged primarily because he himself trespasses against the 'holy spirit of art,' but because he may influence the artist to pervert the 'divine spark' and create popular art."[44] Hoffmann believed that art was the link between everyday reality and an "Other Realm."[45] In "Beethovens Instrumental-Musik,"[46] Hoffmann disparages the artist who disrespects the sacredness of Beethoven's music: "Wer diese Weihe nicht in sich fühlt, wer die heilige Musik zum augenblicklichen Reiz stumpfer Ohren oder zur eignen Ostentation tauglich betrachtet, der bleibe ja davon"[47] [He who does not feel this consecration, who only considers this sacred music as an entertainment, as something to pass the time when there is nothing else to do, as a mere temporary sensuous pleasure for dull ears, or for the benefit of showing himself off—he should leave this music alone].[48] While the film is entertaining, it is more than just a diversion. It is a cinematic creation, a "Gesamtkunstwerk" that combines opera, ballet, cinema, and other visual arts as it invites the viewer into a transcendent world. In fact, it speaks to the desire of the narrator in "Der Sandmann" to present to his readers "a total picture—to strike them with the force of a lightning bolt in his very first word" and it perhaps addresses his fear that the representational power of verbal art cannot compete with that of the visual arts.[49]

Pressburger and Powell's film has seamlessly taken up Hoffmann's ability to provoke a sense of the uncanny through intellectual uncertainty and through the adaptation's ability to create a sense of the familiar and unfamiliar simultaneously. Olympia's status as automaton is strengthened not only through her depiction as a dancing doll but also because of her connection to the filmic apparatus. The 2015 release of the digitally restored film by the Film Foundation and the BFI National Archive indicates the continued popularity of the film.[50] Olympia is alive and well! And, it seems that Offenbach has once again returned from the dead to re-finish his opera. The missing manuscript bought by Jean-Christophe Keck at the auction has motivated a new collaborative effort to produce a definitive edition that would assemble all the known parts in order to create a cohesive understanding of Offenbach's original score and vision.[51]

NOTES

1. William Germano, *BFI Film Classics: The Tales of Hoffmann* (London: Palgrave Macmillan, 2013), 63.
2. Bruce Babington and Peter Evans, "Matters of Life and Death in Powell and Pressburger's *The Tales of Hoffmann,*" in *A Night in at the Opera: Media Representations of Opera,* ed. Jeremy Tambling (London: John Libbey & Company Ltd., 1994), 156.
3. William Germano, *BFI Film Classics*, 20; both "Die Gesellschaft im Keller" and "Das verlorene Spiegelbild" are titles of two sections of Hoffmann's literary tale, "Die Abenteuer der Sylvesternacht," but their details are parceled differently in the film.
4. Heather Hadlock, *Mad Loves: Women and Music in Offenbach's "Les Contes d'Hoffmann"* (Princeton: Princeton University Press, 2014), 42.
5. Gérald Caillat, *The Missing Manuscript: The Story of the Tales of Hoffmann, by Jacques Offenbach*. Documentary. Idéale Audience International, Production Company, and Sept/Arte 2004 (ARTE France). Television.
6. Heather Hadlock, *Mad Loves*, 11–12.
7. Linda Hutcheon, *A Theory of Adaptation* (New York and London: Routledge, 2013), 8–9.
8. William Germano, *BFI Film Classics*, 20.
9. Linda Hutcheon, *A Theory of Adaptation*, 38.
10. Siegfried Kracauer, "Theory of Film," in *Film Theory and Criticism*, ed. by Gerald Mast, Marshall Cohen, and Leo Braudy (New York and Oxford: Oxford University Press, 1992), 250.
11. Siegfried Kracauer, "Theory of Film," 251.
12. Linda Austin, "Elaborations of the Machine: The Automata Ballets," *Modernism/Modernity*, 23, no. 1, (2016): 66.
13. Linda Austin, "Elaborations," 67.
14. Linda Austin, "Elaborations," 67.
15. Christian Metz, "The Imaginary Signifier," in *Film Theory and Criticism*, ed. Gerald Mast, Marshall Cohen, and Leo Braudy (New York and Oxford: Oxford University Press, 1992), 732.
16. Christian Metz, "The Imaginary Signifier," 732.
17. Marcia Citron, *Opera on Screen* (New Haven and London: Yale University Press, 2000), 127–138.
18. Kara Reilly, *Automata and Mimesis on the Stage of Theatre History* (New York: Palgrave MacMillan, 2011), 133–134.
19. Lesley Stern, "An Instance of Operality," in *Between Opera and Cinema,* ed. Jeongwon Joe and Rose Theresa (New York and London: Routledge, 2002), 53–54.
20. Jacques Offenbach. *The Tales of Hoffmann*, performed by the London Symphony Orchestra conducted by Julius Rudel (1972; Hamburg: Deutche Grammophon GmbH), Audio CD.

21. Jean-Louis Baudry, "The Apparatus: Metaphysical Approaches to the Impression of Reality in Cinema," in *Film Theory and Criticism*, ed. Gerald Mast, Marshall Cohen, and Leo Braudy (New York and Oxford: Oxford University Press, 1992), 694.
22. Jean-Louis Baudry, "The Apparatus," 694.
23. In what has become a standard editon of Hoffmann's German text of "Der Sandmann," (E.T.A. Hoffmann, *Sämtliche Werke*, vol. 3, *Nachtstücke, Klein Zaches, Prinzessin Brambilla: Werke 1816–1820,* ed. Hartmut Steinecke w/ collab. Gerhard Allroggen [Frankfurt am Main: Deutscher Klassiker Verlag, 1985 (2009)] [= *DKV* 3]), the editors adhere to the traditional German spelling of the names "Nathanael" and "Olimpia." Various translations and adaptations into French and English have altered the spellings somewhat. Thus, the ballet-opera libretto shows the spellings "Nathaniël" and "Olympia," while the materials for the film spell the characters' names "Nathaniel" and "Olympia." I will employ the spellings "Nathanael" and "Olimpia" in this chapter when addressing the literary tale, and the spellings "Nathaniel" and "Olympia" in reference to the libretto or the film.
24. Kara Reilly, *Automata and Mimesis*, 114.
25. Kara Reilly, *Automata and Mimesis*, 132.
26. Laura Mulvey, "Visual Pleasure and Narrative Cinema," in *Film Theory and Criticism*, ed. Gerald Mast, Marshall Cohen, and Leo Braudy (New York and Oxford: Oxford University Press, 1992), 748.
27. Laura Mulvey, "Visual Pleasure and Narrative Cinema," 753.
28. Kara Reilly, *Automata and Mimesis*, 115.
29. Linda Austin, "Elaborations," 70.
30. Gérald Caillat, *The Missing Manuscript*.
31. See, for example, Wolfgang Preisendanz, *Humor als dichterische Einbildungskraft: Studien zur Erzählkunst des poetischen Realismus* (Munich: Fink, 1976), 47–117.
32. Marcia Citron, *Opera on Screen*, 127.
33. Maria Tatar, "E.T.A. Hoffmann's 'Der Sandmann': Reflection and Romantic Irony," *MLN* 95, no. 3 (1980): 588.
34. Maria Tatar, "E.T.A. Hoffmann's 'Der Sandmann,'" 589.
35. See Frederick Burwick's contribution, "The Uncanny Valley: E.T.A. Hoffmann, Sigmund Freud, Masahiro Mori," in the present volume.
36. Lesley Stern, "An Instance of Operality," 51.
37. Maria Tatar, " E.T.A. Hoffmann's 'Der Sandmann,'" 588.
38. Maria Tatar, " E.T.A. Hoffmann's 'Der Sandmann,'" 607.
39. Jacques Offenbach, *The Tales of Hoffmann*.
40. William Germano, *BFI Film Classics*, 87–88.
41. Bruce Babington and Peter Evans, "Matters of Life and Death in Powell and Pressburger's *The Tales of Hoffmann*," 151.
42. Adam Bresnick, "Prosopoetic Compulsion: Reading the Uncanny in Freud and Hoffmann," *The Germanic Review* 71, no. 2 (1996): 114–132, 116.
43. Adam Bresnick, "Prosopoetic Compulsion," 118.
44. Diana Stone Peters, "E.T.A. Hoffmann: The Conciliatory Satirist," *Monatshefte* 66 no. 1 (1974): 55–73, 58–59.
45. See Kenneth Negus, *E.T.A. Hoffmann's Other World: The Romantic Author and his "New Mythology"* (Philadelphia: University of Pennsylvania Press, 1965).
46. Ernst Theodor Amadeus Hoffmann, *Fantasiestücke in Callot's Manier. Sämtliche Werke in sechs Bänden*, Vol. 2/1, ed. Hartmut Steinecke with collab. of Gerhard Allroggen and Wulf Segebrecht [= *DKV* 2/1] (Frankfurt am Main: Deutscher Klassiker Verlag, 1993), 52–61.
47. *DKV* 2/1, 61.

48. E.T.A. Hoffmann, "Beethoven's Instrumental Music: Translated from E.T.A. Hoffmann's 'Kreisleriana' with an Introductory Note," trans. Arthur Ware Locke, *The Musical Quarterly* 3, no. 1 (1917): 132.
49. Maria Tatar, "E.T.A. Hoffmann's 'Der Sandmann,'" 591.
50. Susan King, "An even more enchanting Tales of Hoffmann after restoration," *Los Angeles Times*, March 11, 2015, http://www.latimes.com/entertainment/movies/la-et-mn-tales-of-hoffman-20150312-story.html.
51. Gérald Caillat, *The Missing Manuscript*.

3

UNCANNY PROSTHETICS

Wounded Bodies in the Lithographs of Théodore Géricault, 1818–1820

PETER ERICKSON

IN 1814, THE FRENCH ROMANTIC painter Théodore Géricault presented a new work at the *Salon* in Paris, an annual exhibition that could make or break the career of a young artist (fig. 3.1). In it he introduced a motif that has long been neglected by scholars but that would become ever more essential for his work: a wounded cavalryman descends on foot down a steep slope. His lean backward is implausible given the way that his horse is simultaneously rearing upward, but this strange and untenable position serves to generate a striking parallel between his left leg, that of the horse, and the saber being employed as a makeshift crutch (fig. 3.2). This parallel draws an analogy between leg, horse, and crutch.

Over the course of his career Géricault turned again and again to the depiction of wounded veterans or disabled panhandlers as they make use of crutches, canes, and artificial limbs. Through the use of these prosthetics, he created unorthodox and even uncanny compositions in which wounded or disabled bodies lean on one another for support.[1] He depicted not only medical prosthetics, such as crutches, wooden legs, and canes, but he also obsessed over the details of uniforms and carriages, as well as the intricate designs of harnesses and reins. It is an aspect of his work that has yet to receive sustained critical attention.

Most remarkably, Géricault's exploration extended to what one might call *animate* or *living* prosthetics: the neighbors on whom we lean for support, the horses on which we ride, and, even more strangely, the use of our own limbs. Géricault's work, I would argue, is profoundly motivated by an interest in the nature of *embodiment*—in the way we experience our limbs as our own, in the way we interact with tools and with others, and in the way we inhabit the world. This interest of Géricault's in the nature of embodied experience contradicts the assumption, present in much of the recent scholarship, that his work is primarily concerned with anxiety and fragmentation.[2]

Figure 3.1 Théodore Géricault, *Wounded Cuirassier Leaving the Field of Battle*, 1814 (Musée de Louvre, Paris)

In the end, it would be difficult to claim that Géricault's complex and varied work presents anything like a philosophical thesis about the nature of embodiment, but all of his paintings continually seem to return to, and try to work out, a central problem that concerned him throughout his brief career. (He died in 1824 in the wake of a riding accident at the age of only thirty-two.)

Figure 3.2 Théodore Géricault, Detail of *Wounded Cuirassier*

The theme of embodiment is woven together in his work with questions of form and pictorial composition: One of the most commonly recurring motifs in the work of Géricault is that of the compound or composite body. Throughout his brief career, and most famously in his equestrian paintings, Géricault seemed to experiment with the possibility of fusing bodies, both of humans and of animals. The human body is, as we shall see, frequently represented in Géricault's work as modular[3]—as composed of a set of interchangeable parts, to be re-arranged and re-assembled at the will of the artist. The original, "natural" unity of the human form is supplanted by the artificial and mechanical unity of the work of art. The classical male nude of the French academic tradition is replaced by the equipped, prostheticized body of the modern era.

Géricault thus offers a re-definition, or reinterpretation of traditional understandings of the human body and its relation to its environment. The classical nude is subjected to a critique not only in terms of its artistic value, but also in terms of its ability to represent the kinds of bodies that, at least by the nineteenth century, had come to inhabit the world. As embodied selves, I believe Géricault wants to argue, we are always already engaged with tools and making use of instruments. Far from merely allegorizing the decline of Napoleonic France or serving as objects of pity, these prosthetics serve a much deeper function: with them, Géricault explored the "absorbed" or pre-conscious relationship of human beings to tools and technology.

This preoccupation with prosthetics is especially evident in Géricault's lithographs. Géricault turned to the recently invented medium of lithography, the latest advance in the mechanical reproduction of images for a mass audience, at the start of the nineteenth century, towards the end of his brief career. It was in these lithographs that he most fully developed his interest in prosthetics: in part because the lithograph forced him to realize his compositional aims on a smaller scale (on the level of the human body) but also because these prosthetics help to thematize the mechanical process of lithographic reproduction, the new dependence of the life and circulation of images on technology.[4]

My goal throughout the essay is to reinterpret and provide context for a set of motifs in Romantic painting—and in Romantic art more generally—that have often been seen as "dark" or "horrific," demonstrating that they were driven instead by a set of interests that go beyond a mere curiosity in the grotesque. My account of Géricault's Romanticism thus centers on his preoccupation with problems of embodiment and of formal and pictorial unity—a set of challenges that, as I will show, confronted painting at this historical moment.

WOUNDED VETERANS

In her *Extremities: Painting Empire in Post-Revolutionary France*, Darcy Grigsby focuses on the social and political promise offered by the compound bodies Géricault depicted in his images of wounded veterans. Géricault, Grigsby writes, showed a penchant for depicting "monumental mounds of interdependent bodies."[5] For Grigsby, the lithograph of a cart of wounded soldiers seems to enact, both literally and corporeally, new possibilities for human community in a world conceived without social or political hierarchies (fig. 3.3). According to Grigsby: "Rather than the disciplined, organizing structure of military rank, physical disabilities—so many random inflictions—now determine spatial relationships among men. In these images, the man who can sit holds the man with the head wound who lies in his lap; the man who can stand with a crutch holds onto the cart; the armless man leads the blind man; the man capable of walking carries the man who cannot."[6]

Figure 3.3 Théodore Géricault, *Cart Laden with Wounded Soldiers*, 1818 (Art Institute, Chicago)

In these images, "social cohesion" is produced, as Grigsby puts it, "through the complementarity of differently wounded men."[7] Bodily injury provides the foundation for new forms of political community in the wake of the French Revolution and of Napoleon's defeat on the battlefield. In Géricault's most famous painting, *The Raft of the Medusa*, the shipwrecked survivors of a French military frigate have been stripped of their military rank, stripped indeed of their uniforms, and form a kind of human pyramid, leaning towards a sail glimpsed on the horizon.

Grigsby's description of these wounded bodies was crucial for the development of my ideas about Géricault's work. In my essay here, though, I want to focus on the formal and aesthetic reasons why Géricault was so interested in compound bodies. This is not to downplay the political and psychological content of these images but rather to emphasize that this struggle in Géricault's work, to produce unity out of an array of differently wounded bodies, also had formal and aesthetic motivations.

In his 1818 lithograph, *The Return from Russia*, Géricault depicts two survivors of Napoleon's invasion in the foreground, working their way home through the tundra (fig. 3.4). From his sketches and preparatory drawings, we can see just

Figure 3.4 Théodore Géricault, *The Return from Russia*, 1818 (Art Institute, Chicago)

how carefully composed (and even contrived) this image is. Géricault seemed to experiment with precisely which wounds he should inflict in order to render the two figures ever more dependent on each other. He wondered whether the soldier on the horse should be literally or figuratively blinded—either with a bandage across his face or with a traumatized, vacant look in his eyes (fig. 3.5). And Géricault experimented with various arrays of crutches and canes (fig. 3.6). In one sketch, he actually imagined two soldiers, each with one crutch, hobbling along arm in arm (fig. 3.7). (Note again on the sketchbook page the visual analogy between the crutch and a horse's leg in fig. 3.7.)

In the end, for the final lithograph, he decided on a pair in which their opposing limbs have literally been removed or at least rendered inoperative. (The rider on the horse is missing his left arm, and the soldier walking beside him is missing his right [fig. 3.4].) Any limb that does not participate in this relation of mutual support has been excised in order that we not doubt the *necessity* of this composition and of this particular organization of figures. This theme is redoubled through the addition of a second pair of soldiers in the background in which one figure is carrying the other.

Géricault sought to create constellations of figures that would seem necessary, that would gain a certain urgency, and that would thus heighten our sense of the unity of his compositions. He uses wounded bodies, with their mutual dependence and their reliance on prosthetics, to generate our emphatic conviction in the unity of the *tableau*. The wounded veterans manifest relations of support that allow the arbitrary arrangement of figures on the canvas (or, in this case, on the page) to seem anything but. We experience a sense of conviction that the figures could not have been arranged any other way.[8] Géricault ultimately explored a range of variations on this theme: in another drawing (*Dappled Draught Horse Being Shod*), for example, instead of the horse serving as the prosthetic support of the man, the man could serve as the temporary crutch of the horse (fig. 3.8).

In an uncanny way, Géricault's images create compound bodies, in which the means of motion, sense perception, and support are shared. In *Return from Russia*, it is not the soldiers—who have been dismembered and traumatized—but ultimately the horse that carries the burden of agency and seems to be tracing its way home through the tundra (fig. 3.4). Géricault provides us with images of compound or composite bodies in which horse and rider, or human being and prosthetic, have been rendered inseparable or even seem to have been fused together.

The heightened sense of unity manifested by this group is redoubled in the way that there is only the barest indication of a world outside of this cluster of

Figure 3.5 Théodore Géricault, *Study of Figure Group*, with Detail of Grenadier's Head, 1818 (Art Institute, Chicago)

Figure 3.6 Théodore Géricault, *The Retreat from Russia*, Watercolor, 1818 (Musée des Beaux Arts, Rouen)

bodies. They exist in an indeterminate space—defined not with respect to their surroundings and context but only in relation to one another, indeed even exclusively in terms of their mutual *dependence* on one another.[9] It is a unity in which no part, no figure, is unnecessary, and in which each serves some kind of function for the whole.

Figure 3.7 Théodore Géricault, *Studies of Heads and Two Compositional Studies*, 1818 (Art Institute, Chicago)

Figure 3.8 Théodore Géricault, *Dappled Draught Horse Being Shod*, ca. 1823 (Museum of Fine Arts, Boston)

THE DEMANDS OF PICTORIAL UNITY

Recent studies have emphasized the historicity of the seemingly timeless idea of pictorial unity.[10] Notions of unity depend heavily on genre and are closely related to a set of norms and conventions that have shifted over time. The unity of a composition has been understood in a variety of ways and been subject to varying levels of emphasis. Thomas Puttfarken, in *The Discovery of Pictorial Composition*, traced the origins of modern theories of pictorial composition to France in the seventeenth century. This new emphasis on composition in France marked a shift away from pictorial values that had been dominant in the Italian Renaissance, such as figural presence, volume, and spatial perspective.[11] Influenced by Aristotelian theories of tragedy, early modern French theorists of composition emphasized the representation of dramatic unity in which the depicted figures seemed to be participating in or reacting to a common action.[12]

Decisions as to lighting and the distribution of figures on the canvas should be directed, it was argued, towards rendering the action comprehensible, plausible, and legible—making clear, for example, which are the principal figures and what their relationship to one another is—such that the viewer is able to understand the moral import of the painting and benefit from it. "Formal devices," Puttfarken writes, "were not acknowledged as having their own interest, they were treated purely as means of directing our eyes to the crux of the action, of concentrating our attention on the subject-matter."[13] Paintings were organized visually primarily in order to render their didactic message the more intelligible to their audience.[14]

The seventeenth and eighteenth centuries also witnessed a shift in the function and audience of painting. If Renaissance frescos and altarpieces, for example, were embedded in architectural features and had sought to depict life-sized figures who seemed to inhabit our space, to face us, and to exhibit a certain presence and authority, the *tableau* (or mid-sized easel painting) that became popular in the seventeenth century had an altogether different relationship with the beholder. Hung on the wall of a gallery, *tableau* paintings were more easily portable, more self-contained, more modest in size, and more likely to emphasize relationships *within* the depicted world of the painting, rather than with the beholder or with the (now arbitrary) architectural features surrounding it.[15] They worked in an altogether different context and emphasized different pictorial values.

By the middle of the eighteenth century, pictorial unity (based originally, as I mentioned above, on the unity of action) had acquired a new level of urgency. In his landmark 1980 study, *Absorption and Theatricality*, the art historian Michael Fried identified a seemingly obsessive effort in late eighteenth and early nineteenth century French painting to evade charges of theatricality.[16] Painters were praised by critics to the extent that they depicted figures who appeared to be unaware of being observed. In this sense, artists sought to create paintings that seemed to deny the presence of a beholder before the canvas—as if acknowledging the presence of a viewer before the painting would fundamentally concede that the depicted figures were merely posed and hence unworthy of our interest.

Pictorial unity thus came to take on new meaning. It served not only to ensure the painting's legibility, as it had for French theorists in the seventeenth century, but also to provide the painting with a sense of *closure*. It served to close off the depicted world of the painting from that of the beholder, to insist on its own integrity, and thus to prove that the painting was not merely posed or staged for the beholder's benefit. The problem—and this was, according to Fried, what drove the development of French painting in the eighteenth century—was that every convention that had previously served to assert the authenticity of the figures depicted in the painting came to be seen as itself too obviously directed at the beholder. Denying the presence of the beholder required ever more elaborate and convoluted arrangements of figures. The dramatic action represented in the painting needed to be ever more intense, such that the figures in the painting could not possibly be distracted from it, could not possibly betray an awareness of a spectator.[17] (I return to this problem below.)

The link between pictorial and *corporeal* unity is an old one. In the Renaissance, the term "composition" indeed referred not to the overall structure of an image—in the sense in which we use the term today—but merely to the successful rendering of the human body.[18] Composition was thus understood not as a quality of the overall painting but rather of the body being represented. The various limbs and members of the human body were expected to be proportionate to one another. And the parts of the body were supposed to be unified in terms of action. To quote Puttfarken, characterizing Leone Battista Alberti's view: "If a man is running, then not only his feet must be shown moving but also his hands. In a dead man there must not be a member that does not seem completely lifeless, whereas in a living figure all members must seem completely alive, and that means they must be engaged in their appropriate movements."[19] Thus Leonardo da Vinci could argue, in his notebooks, for the superiority of painting over poetry on the grounds that: "There is the same difference between the poet's and the painter's representations of the human figure as there is between dismembered and united bodies. Because the poet in describing the beauty or ugliness of any figure *can only show it to you consecutively, bit by bit,* while the painter can display it all at once."[20] The poet, relating to us the beauty of one body part after another, can only ever offer us at best a dismembered corpse, whereas the painter can present the body as a whole.

It was Shaftesbury who, in the eighteenth century, first connected the theme of corporeal unity, metaphorically, to the *tableau* or easel painting.[21] The *tableau* should not be defined merely by "the Shape or Dimension of a Cloth or Board," but by its insistence on unity—a unity that is explicitly related to that of the body. The *tableau*, according to Shaftesbury, is "a *Single Piece*, comprehended in one *View*, and form'd according to one single Intelligence, Meaning, or Design; which constitutes a real *Whole*, by a mutual and necessary Relation of its Parts, the same as of the Members in a natural Body."[22] Shaftesbury's reinterpretation of the traditional relationship between composition and corporeal unity reflects an escalating insistence on pictorial unity. Instead of merely being concerned with the proper depiction of the body, pictorial composition should now aspire to the same level of necessity as the natural body. It should insist, indeed, on a kind of *organicity*.

With Géricault, this compulsion to assert the unity of pictorial composition arrives at a new extreme: in his paintings and lithographs, Géricault frequently asserts the unity of his compositions *over and against* the unity of the body. He asserts their unity, in other words, not through depicting the "natural," male, academic nude but, as in his notorious oil sketches of amputated limbs, through explicitly disassembling and reassembling the body into new configurations. According to Linda Nochlin, in these paintings, "the coherence of the body is totally shattered. The dispersed fragments are then reconjoined at the will of the artist in arrangements both horrific and elegant."[23] For Nochlin: "Géricault consigns the mutilated heads to the realm of the object, plays their erstwhile role as the most significant part of the human body against their present condition as lifeless, gruesome fragments, deployed on a tabletop like meat on a butcher's counter, or specimens on a dissecting table. Yet, even more disturbingly, the heads have been arranged here for maximum effect by the controlling artist: Géricault's project here is an aesthetic one, involving formal intervention."[24] What is so striking about these images of amputated limbs, however, is that these fingers and limbs seem so eminently capable of touch. They even, as more than one art historian has noted, seem to be locked in an erotic embrace.

One finds again and again in the scholarship the claim that Géricault's work is primarily about fragmentation. Precisely the uncanny thing about Géricault's images, however, is the way that the fragmented bodies in his compositions seem to come together. In another painting, the heads of two beheaded criminals are arranged on what appears to be a marital bed, and they are so strikingly related to each other that one would be unsurprised for them to break into pillow talk (fig. 3.9).[25]

Géricault disassembles the body in his work and reassembles it in new configurations. These new constellations are not so much political, however, as they are a tribute to the power of painting. Bodies are disassembled, reassembled, and reanimated in compositions that have a real urgency and poignancy. What we find in Géricault's work are not so much fragments but new forms of compound bodies and new possibilities for pictorial composition, which are in turn—crucially—revelatory about new possibilities for human existence.

It can hardly be overstated how far we have come here from the tradition of the French academic nude. The unity of the painting is realized in a form of corporeal unity. It is not, however, the unity of the natural body but rather a unity composed through the work of the painter. The disabled and prostheticized body, which plays such a role in Géricault's work, insofar as it seems to be both more and less than an "ordinary" body, unsettles normative understandings of the "wholeness" and integrity of the body.[26] James Porter, among others, has argued that the disabled body both exceeds and falls short of conventional expectations. On the one hand, the disabled body "appears to lack something essential . . . ; it seems *too little* a body: . . . not quite a body in the full sense of the word, *not real enough*." On the other hand, however, the disabled body appears as "somehow *too much* a body, *too real*, too corporeal: it is a body that, so to speak, stands in its

Figure 3.9 Théodore Géricault, *The Severed Heads*, ca. 1818–1820 (National Museum, Stockholm)

own way."[27] The disabled body in Géricault's work, both excessive and deficient, challenges how we think of organic unity.[28]

Tobin Siebers has gone so far as to argue that disability not only provided an essential subject matter for modern art, but that modern art was commonly itself associated with disability (for example, see the frequent criticisms of modernism as "entartet"). "Specifically, the Nazis," Siebers notes, "rejected the modern in art as degenerate and ugly because they viewed it as representing physical and mental disability."[29] Modern art, according to Siebers, not only frequently sought out representations of disability, but it identified disability as a crucial set of resources for thinking through fragmentation, creativity, and non-normativity. Géricault might count, in this sense, as a precursor to the work Siebers describes, but he also does so in a creative and unexpected way. The aestheticizing of disability is not without its pitfalls. It can serve to efface—and to substitute itself for—the real lived experiences of the disabled. But it can also, as I have shown, serve to heighten our sense of the prostheticized and augmented nature of all human existence and of the nature of embodied subjectivity.

GÉRICAULT AND THE PROBLEM OF THEATRICALITY

I mentioned above how figurative easel paintings in the gallery, because of the way that they faced viewers, came to appear theatrical and posed. According to Fried, Géricault made use of a variety of strategies in his attempt to evade this charge of theatricality and to avoid having his figures appear posed. First, especially in his early, intense work, Géricault sought to depict figures that "aspired to go beyond the theatrical by virtue of the *sheer excessiveness* of their commitment" (my emphasis). It is a sign of just how far Géricault pushed this dynamic that his figures, Fried writes, so often appear to be committed to nothing other than "their own exalted or excruciated physicality."[30] These paintings and drawings seem to have no content other than the "tormented physicality of the antagonists' bodies."[31]

The intention is to achieve such dramatic and physical intensity that it would no longer be possible to reflect on the presence of an observer. Géricault seeks to efface or at least obscure the presence of self-conscious posturing. He strove in this way for an ever more intense physical immediacy, unimpeded by self-awareness. In *Mameluck défendant un trompette blessé* (fig. 3.10), we have an example of just such a hyperbolically charged situation in which the instability of the ground is emphasized and the picture plane itself is threatened with rupture.

The second strategy that Géricault deployed was to displace or delegate the depiction of inwardness to another agent, one considered naturally antitheatrical. Horses, and animals generally, seem to be incapable of *posing.* They carry out the roles assigned them (pulling a wagon or bearing a load, for example). They appear to be incapable, in other words, of anything other than total involvement in their own physical activity.[32] Characteristic of Géricault is the depiction of horse and rider in which the rider has been figuratively or even literally blinded, and the horse has assumed the task of expressivity—as it were, on behalf of the rider. In one painting, the horse actually becomes the subject of a human portrait (fig. 3.11).

Perhaps, most beautifully, there is a wash drawing in which the owner of a horse has fallen asleep, and his horse gazes fretfully on his behalf into the looming darkness (fig. 3.12). The horse seems to keep watch *on behalf of* its owner. (One wonders if a more finished version of the drawing, in further differentiating between the figures, would have diminished this impression of shared perception.) In many of Géricault's lithographs as well, the burden of agency has been displaced from one figure to another, as we have seen in the case of *Return from Russia*. The sol-

Figure 3.10 Théodore Géricault, *Mameluke Defending a Wounded Trumpeter*, 1818 (Metropolitan Museum of Art, New York)

dier who can see does so for the soldier who has been blinded. The amputee leans on the soldier who can feel out the ground ahead of them. This sharing of sensory perception renders these figures more convincing in rendering them ever more dependent on one another, but it also raises questions about the nature of sensual perception and the existence of a shared world.

Figure 3.11 Théodore Géricault, *Head of a White Horse* (Musée de Louvre, Paris)

These two techniques identified by Fried—the attempted effacement, on the one hand, and the displacement or delegation of consciousness, on the other—are not entirely distinct from one another but constitute instead two poles of an axis along which much of Géricault's most ambitious early work can be arrayed. Géricault's interest in prosthetics, it seems to me, constituted a third strategy, one closely intertwined with the others but nevertheless distinct and worthy of attention.

Figure 3.12 Théodore Géricault, *Two Draft Horses with a Sleeping Driver*, 1820–1822 (Metropolitan Museum of Art, New York)

Géricault's interest in prosthetics has, moreover, as I mentioned at the outset, everything to do with his engagement with the new medium of lithography.

Géricault turned to lithographs, then a relatively new method of producing prints, during a period when he was frustrated in his ambitions to produce large-scale history paintings. His *Raft of the Medusa* (1819), a painting that was more than twenty-three feet wide and over sixteen feet tall, was a popular success but nevertheless met with mixed reviews from critics.[33] Compelled to realize his ambitions in a smaller format, that of the lithograph, Géricault turned to new pictorial configurations. If the *Raft of the Medusa* had constructed a massive human pyramid of straining, multi-racial bodies, Géricault's lithographs were characterized, on the other hand, by a really extraordinary compression: Géricault attempted to realize the aims of history painting in a smaller format, in much tighter compositions, and on a smaller scale. (The print reproductions here give no sense of the extreme difference in scale between a massive history painting and a fourteen-by-seventeen-inch lithograph.) Géricault strove to recreate the effect of his multi-figure history paintings on the level of the human body—in figures that were a hybrid of bodies and prosthetics. If, in the French academic tradition, the male nude had been considered sacrosanct, Géricault's paintings and lithographs represented nothing less than the de-privileging of the human body as the fundamental, indivisible unit of history painting. It is as if the charge, present from the seventeenth century onward in French painting, to represent only the most essential figures, to include nothing in the painting that was unnecessary and that might detract from the painting's composition, had ultimately resulted in a form of composition on the level of the body itself.

Prosthetics had for Géricault, moreover, certain physical limitations that made them all the more compelling for pictorial composition. They stand in for legs but nevertheless have a different range of motion than legs do. The artificial limb necessitates a level of frontality that would otherwise be difficult to justify. Whereas the figures in Géricault's early compositions are often contorted and display an exaggerated level of torsion in order to demonstrate their commitment to their activity, here—in Géricault's *The Swiss Sentry at the Louvre* (fig. 3.13)—the presentation is remarkably direct. The wounded veteran can face directly out at the picture plane, without seeming forced or theatrical.

In this scene, a Swiss mercenary guard at first tries to stop an amputee, whom he takes for a beggar, from entering the Tuileries. The amputee responds by defiantly pulling back his lapel to reveal a medal of the Legion of Honor, indicating that he is a veteran of the Napoleonic wars and compelling the Swiss mercenary to salute him. The composition alludes to the forceful way that the figures had planted themselves in Jacques-Louis David's *Oath of the Horatii* (fig. 3.14). The veteran has taken on a heroic stance, that of a classical hero, but it is made possible by, and it is emphasized by, his use of a prosthetic. The force of his stance is further reinforced by the fact that the Swiss guard is being pushed back into an architectural feature on the right, which gives the composition, otherwise taking place on an open street, a sense of constrained space.[34]

THE UNCANNY AND EMBODIMENT

I have argued above that Géricault turned to wounded bodies and prosthetic limbs not to allegorize the decline of post-Napoleonic France or to thematize fragmentation but instead to emphasize the pictorial unity of his compositions. I want to push my argument further by suggesting that, through his exploration of the ways in which the body can be physically supplemented, Géricault further explored—on a deeper level—the nature of embodied experience, the ways we interact with things and objects, and the ways we inhabit the world. I will do so here, counter-

Figure 3.13 Théodore Géricault, *The Swiss Sentry at the Louvre*, 1819 (Art Institute, Chicago)

intuitively, by engaging with the German Romantic concept of the uncanny (*das Unheimliche*)—not so much as a psychological category but rather as a *formal* strategy, a certain way of manifesting unity.

To demonstrate this dependence on tools and prosthetics, Géricault is attracted to extreme situations where the use of prosthetics is dangerously close to

Figure 3.14 Detail of Jacques-Louis David, *The Oath of the Horatii*, 1784 (Musée de Louvre, Paris)

rupture or where the technical apparatus is so complex that it verges on the implausible or even the bizarre. In this way, he can demonstrate just how intensely and unreflectively the physical implement is being employed. The prosthesis is rendered all the more apparent to the spectator, without necessarily becoming conspicuous to the figures in the painting themselves. In *The Swiss Sentry at the Louvre* (fig. 3.13), this is achieved by highlighting the continuity in the image between the artificial limb, the crutch, the rifle and the architectural support of the column, which together form a kind of W-shape. To my mind this serves again to draw a kind of visual analogy between leg, cane, crutch, and architectural support, but it also draws the beholder's attention to the prosthetic as an essential feature of the tableau. In this way, Géricault is able to render prostheses present as a mode of support in the midst of their unreflective use, without having to render them inoperative or obtrusive, in the Heideggerian sense.[35]

This can be seen most clearly in Géricault's portrait of the young Alfred Dedreux (fig. 3.15). The beige sleeve of the young boy's cloak appears to have grown into a third arm—an appendage, that is, if anything more muscular and more robust than his other limbs. Dedreux's clothes seem, in an uncanny and unset-

Figure 3.15 Théodore Géricault, *Portrait of Alfred Dedreux as a Child*, ca. 1819–1820 (Metropolitan Museum of Art, New York)

tling way, to have become an extension of himself, rendered even more vivid by the way that the beads of his trousers sparkle as if they were so many additional eyes and the way that his clothing seems to billow out around his midsection. Trying to find the boundary here between boy and cloak is curiously, unsettlingly impossible. It seems at least possible that the boy is intentionally staging the scene as a prank. By intentionally removing his arm from the sleeve of his cloak, he can adjust his posture to provoke uncertainty in the spectator, as if this were all a trick on the playground used to terrify his classmates. In this way, the boy demonstrates his mastery over these extensions of his body and a keen awareness of how they are perceived by others. Much like the artist Géricault himself, the boy challenges the viewer's attempt to make a distinction between the body and its accessories.

It is not, therefore, that we learn to distinguish between the mechanical and the organic in viewing these images, that we somehow get better at distinguishing between real and artificial limbs, that we become more adept at identifying them, but rather, on the contrary, that Géricault's work allows us to realize how the body is itself mechanical, always already supplemented, and engaged with a world. In his use of composite groups of bodies and implements, Géricault demonstrates that individual bodies are themselves also composites. He makes manifest the fundamentally mechanical, the acutely immanent, nature of the human body—an understanding that we might find self-evident or self-explanatory but that was, in some sense, still revolutionary for the eighteenth and early nineteenth centuries.

The German psychologist Ernst Jentsch in his deeply influential 1906 essay "On the Psychology of the Uncanny"—so often overshadowed by Sigmund Freud's essay on the same topic—describes how viewing the mentally ill, and in particular witnessing an epileptic fit, forces us to recognize the mechanical processes underpinning what we had always understood to be a transcendent whole: "[T]he dark knowledge dawns on the unschooled observer that mechanical processes are taking place in that which he was previously used to regarding as a unified psyche. [. . .] [T]he epileptic attack of spasms reveals the human body to the viewer—the body that under normal conditions is so meaningful, expedient and unitary, functioning according to the directions of his consciousness—as an immensely complicated and delicate mechanism."[36] We are so accustomed to viewing the human being as a transcendent subject, endowed with a soul and a coherent inner life, that we experience a deep sense of unease in witnessing an epileptic seizure and we are forced to recognize that the body is a mechanical apparatus, subject to rules of support and gravity. Géricault, incidentally, created a wash drawing during his visit to Italy in 1817 that seems to show just such an analogy between a man in an epileptic fit and a man with a crutch, as if achieving an analogy between them—the man with epilepsy and the man with the crutch—were the purpose of the sketch. The contrast between the physicality of the epileptic man's convulsions and the transcendent nature of his soul is illustrated by the unearthly glow that seems to emanate from him in the drawing.[37]

In his essay on Géricault entitled "Pleasurable Fear," the art historian Stefan Germer devotes significant attention to Géricault's interest in physical supplements. But for Germer, "the hyper-bodies created through the fusion of horses and riders" always demonstrate a profound sense of lack, an anxiety about fragmentation. These "centaur fantasies," as Germer calls them, offer us "fantasies of bodily totality and impregnability, where the fear of fragmentation is met with by an armoring of the body." Géricault strove, it seems to Germer, to defuse this anxiety through "the piling-up of ever more musculature and the ever more deceptive interlacing of horse and rider."[38] Fragmentation, it would seem, should be preempted hyperbolically by grasping all the more tightly and excessively at the outside world.

In contrast to Stefan Germer, I would like to suggest a far more profound significance for the use of prosthetics and tools in Géricault. It is not so much a matter of the distraught masculine subject armoring himself up in the face of existential despair, although I would not dispute that this anxiety is sometimes present in Géricault's work. Rather, these prosthetics demonstrate an intense interest in the ways in which bodies themselves are constituted, in the way tools and instruments seems to become extensions of ourselves, and in the way that the body is itself modular, a composite, a mechanical apparatus of many parts.

The ballet dancer of the German Romantic writer Heinrich von Kleist's famous essay "On the Puppet Theater" even went so far as to argue that reflection creates an awkward distance, a disjunction between subjects and their physical activity. *Grace*, on the contrary, could only emerge from a state of total absorption in one's activity—succumbing to the mechanical laws of motion like a pendulum. "We see," Kleist writes, "that to the degree that, in the organic world, reflection becomes dimmer and weaker, grace shines forth ever more radiantly and more forcefully."[39] If we truly wish to see examples of grace, the ballet dancer argues, then we must turn to the world of puppets, marionettes, mechanical limbs, and automata. In this way, the twin themes of embodiment and antitheatricality intersect. The limbs of *puppets* follow purely mechanical laws of gravity and motion and hence lack this human element of distraction. They do not reflect on the role they are performing; they do not perform for the sake of an audience; they merely engage in mechanical motion. Even the sophisticated prosthetic limbs then being introduced for the wounded in England are described as offering a kind of grace to humankind unknown to real human limbs. Ballet dancers can never overcome this inescapable estrangement and detachment from their own activity. In the end, despite all of their skill, ballet dancers can never wholly give themselves over to the dance. Reflection always creates a certain unavoidable clumsiness.

GENDER AND THE WOUNDED BODY

In closing, it would be remiss of me not to note that the depiction of prosthetics in Géricault is acutely gendered. While the male use of prosthetics is often depicted as heroic—in *The Swiss Sentry at the Louvre*, as we have seen, a wounded veteran is even cheered on by spectators—the women that Géricault depicts making use of prosthetics are often objects of disgust and ridicule. We are even presented, in several of these images, with spectators who react with horror and seem to model our reaction for us. Why is the form of embodied subjectivity, enlivened by the use of tools and instruments, that Géricault presents us with in his lithographs seemingly available only to men? The only defense I can offer is to say that an alternative reading of them is possible.

In one sketch, believed to have been made in Italy, Géricault represents the confrontation between a disabled woman and two young children, presumably on the street (fig. 3.16). The woman, leaning on a cane, has spun about and is gesticulating angrily. Her attitude is defiant and forceful, even if she seems off-balance and her resistance seems futile. The woman's features are androgynous, even distinctly masculine, almost as if she were dressed in drag. Her figure and facial fea-

Figure 3.16 Théodore Géricault, *Scene from Italian Street Life*, ca. 1816–1817 (Collection of Andrea Woodner, New York)

Figure 3.17 Théodore Géricault, *The Artillery Caisson*, 1818 (Art Institute, Chicago)

tures are undeniably represented as monstrous.[40] And the young girl on the right, looking back, seems to make a gesture with her fingers as if to ward off a curse. The sketch is undeniably an astonishing technical achievement. One gets an acute sense from the drawing of the woman's physicality and her manner of motion. Even if it seems to me that she has been sketched with some sympathy, though, I still find the image troubling.

One senses a connection to *The Swiss Sentry at the Louvre*, a lithograph which may have been made in the same year. And the woman's pose, her angry glance backward, her gesture with a clenched fist, and even her uneven footing are, moreover, nearly identical to that of the grenadier in *The Artillery Caisson* from 1818 (fig. 3.17). The resemblance is so striking that I am almost prepared to call the drawing a sketch for the later lithograph, the gender and age of the protagonist notwithstanding. Much like *The Swiss Sentry at the Louvre*, the woman with a crutch here assumes the defiant bearing of a military hero. It is almost as if Géricault tried to force this masculine posture onto the figure of the old woman, only later to give up, finding it too dissonant, and instead grant it to a male grenadier.

Figure 3.18 Théodore Géricault, *A Paraleytic Woman*, 1821 (Art Institute, Chicago)

A Paraleytic Woman (fig. 3.18), a lithograph that Géricault made based on his visit to London in 1820–1821, is a more complex image. In the foreground, a disheveled peasant with a misshapen cap[41] is pulling a strange cart in which a paralyzed older woman is seated.[42] Her passivity is emphasized by the way her cloak has clearly been folded over the edges of the cart by someone else's hands.

The cart itself consists of two wheels attached to a large chair, which has been strapped directly to the back of the peasant, rendering the two inseparable. As in several of the other lithographs of wounded veterans, the piling up of injured bodies serves to create a composite body in which the means of motion, support, and sense perception are shared. In this case, Géricault seems to have been especially interested in how this particular compound body has become a creature with eyes, so to speak, on both sides of its head.[43]

On the left, a young woman is leading a girl away from the scene. The little girl, on the one hand, whose face is completely hidden from the viewer by a small hat, seems to be unaware of the paralytic woman behind her and clutches a toy horse under her arm. The young woman, on the other hand, on the threshold of adulthood, turns around and looks back. In this way, as some have noted, the arrival at the sexual maturity of adolescence corresponds to the onset of a visual maturity.[44] Still on this threshold, the young woman is disturbed by what she sees, glaring backward while simultaneously preventing the little girl with the toy horse from doing the same.

To an astonishing extent, art historians have been so preoccupied with the torn poster on the wall in the background that they have neglected a more comprehensive reading of the composition. The lithograph shows us, in fact, three analogous pairs of figures on a crowded urban street: the young girl and her toy horse on the left; the brutish peasant and the paralyzed woman in her wheelchair in the center; and the horse and carriage on the right.

Each of these three pairs thematizes the relationship between horse and rider and the mutual relations of support and dependency that such a relationship relies upon. We might even arrange these pairs of figures along two axes: on the one hand, from the left, we have the toy horse, the brutish peasant playing the part of a horse, and then the live horse, in a trajectory of ever increasing vitality. On the other hand, again reading left to right, we see the young girl, the paralyzed old woman, and finally the inanimate carriage, in an opposing trajectory of increasing paralysis. One art historian has even suggested that the carriage resembles a funeral hearse, which would heighten the emphasis on progressive dis-animation.[45] It is striking to note that these two axes are also gendered. We are presented with what seems to be an allegory of the development of a young girl from adolescence into old age, paralysis, and, finally, death. This trajectory of steady paralysis bears an inverse relation to the development of the horse across the three pairs. The inanimate toy horse develops into the brutish peasant who at last develops into the fully animated horse pulling the carriage. The development of the young girl into adolescence, maturity, and paralysis is paired with the descent of a second figure, coded male, into bestiality.

This set of axes helps us to locate this strange compound body in the center of the lithograph—the woman who, in her state of paralysis, is half-animate; and the peasant who is half-man, half-horse. But it is also helps us to locate the motif of prosthetics in Géricault's work at the center of a set of inverse relations between animation and paralysis—with a development towards spontaneity, liveliness, and bestiality, on the one hand, and towards death, paralysis, and dis-animation, on the other. These two axes present the two extreme possibilities of either inanimate doll or animate beast. At the hinge, at the pivotal point, at the moment of decision, between these two extremes, there is the young adolescent woman, involved in none of the pairs, seeking to protect the young girl in her care but turning back herself to look over her shoulder. And the lithograph allegorizes her fall from grace, her growing sexual awareness, her visual maturity, and hence the future that awaits her in paralysis, trauma, and death.

The adolescent woman occupies a crucial position in the lithograph, but we should not identify too quickly with her scorn. If we have been attentive to Géricault's work, then we have learned to be fascinated by the artificial contraption in the foreground, to ponder over the details of its functioning, and to marvel at its potential for artistic composition. This, then, is not merely an allegory of lost innocence and sexual awareness, as Linda Nochlin would have it, but also an image about the place of prosthetics in the center of this field of competing forces—a center offering a set of possibilities for human existence but also for art.[46]

NOTES

1. This discussion of prosthetics is deeply indebted to Michael Fried's *Menzel's Realism* (New Haven: Yale University Press, 2002), 100–104. See also Norman Bryson's immensely suggestive reading of the shadow of what appears to be a wooden leg in Géricault's *Seated Hussar Trumpeter* in Bryson's "Géricault and Masculinity," in *Visual culture: Images and interpretations*, ed. Norman Bryson, Michael Ann Holly, and Keith Moxey (Hanover: Wesleyan University Press, 1994), 228–260, 241; and Drew Leder, *The Absent Body* (Chicago: University of Chicago Press, 1990). See, however, David Mitchell and Sharon Snyder's warning against the tendency for prostheses in painting to tend towards allegory in *Narrative Prosthetics* (Ann Arbor: University of Michigan Press, 2001).
2. For an example of such an approach to Géricault's work, see Stefan Germer, "Pleasurable Fear: Géricault and uncanny trends at the opening of the nineteenth century," *Art History* 22, no. 2 (June 1999), 159–183.
3. See the chapter "Modular Bodies" in Stefani Engelstein, *Anxious Anatomy: The Conception of the Human Form in Literary and Naturalist Discourse* (Albany: State University of New York Press, 2008), 113–144, which deals with English and German Romanticism. See also Stefani Engelstein's contribution to this volume.
4. Not least because the lithographs are mostly without color, Charles Clément, in his well-known biography of Géricault from the nineteenth century, believed they were especially revelatory about Géricault's skill at composition: Charles Clément, *Gericault: étude biographique et critique* (Paris: Didier & Cie, 1879), 207. The lithographs are, in my view, like experimental compositional studies.
5. Darcy Grimaldo Grigsby, *Extremities: Painting Empire in Post-Revolutionary France* (New Haven: Yale University Press, 2002), 181.
6. Grigsby, 181.
7. Grigsby, 182.
8. In this way, as I show below, Géricault embodies certain ideas of the eighteenth-century French theorist Denis Diderot who wrote: "A composition must be organized so as to *persuade me* that it could not be organized otherwise; a figure must act or rest so as to persuade me that it could not do otherwise." Cited in Michael Fried, *Absorption and Theatricality* (Chicago: University of Chicago Press, 1988), 85–86 (my emphasis). Fried comments on Diderot: "The question he seems always to have asked himself is not whether a particular painting could be shown to possess an internal rationale that justified and in that sense bound [it] together . . . but whether his actual experience of the painting, prior to any conscious act of reflection or analysis, persuaded him beyond all doubt of the work's dramatic and expressive unity" (85).
9. Fried points to the way that Géricault's *Charging Chasseur* has been "depicted in a manner that isolates them from their surroundings, or at least doesn't make the meaning of their

actions dependent either on those surrounding or on a framing *récit* in any specific way." But Fried sees these isolated figures "less as a fragment than *as a new sort of unit or gestalt*" (my emphasis). This point is crucial for my own argument. See Fried, "Géricault's Romanticism," in *Géricault*, ed. Régis Michel (Paris: Documentation Française, 1996), 647.

10. Along with Puttfarken's study, which I discuss at length here, see Hans Körner, *Auf der Suche nach der 'wahren Einheit': Ganzheitsvorstellungen in der französischen Malerei und Kunstliteratur vom mittleren 17. bis zum mittleren 19. Jahrhundert* (Munich: Wilhelm Fink Verlag, 1988).
11. Thomas Puttfarken, *The Discovery of Pictorial Composition: Theories of Visual Order in Painting 1400–1800* (New Haven: Yale University Press, 2000).
12. Puttfarken, *The Discovery of Pictorial Composition,* 236–239. See also the chapter "Félibien and the Early Academy: *Unité de Sujet* and *Convenance*" in Puttfarken, *Roger de Piles' Theory of Art* (New Haven: Yale University Press, 1985), 1–37; and Puttfarken, "David's *Brutus* and Theories of Pictorial Unity in France," *Art History* 4 no. 3 (September 1981), 291–304.
13. Puttfarken, *Roger de Piles' Theory of Art*, 131.
14. An exception to this trend, according to Puttfarken, is the work of Roger de Piles, a French theorist who in the seventeenth century elaborated an alternative model of the unity of the *tout-ensemble* of the composition that relied on colorism and the artist's success in creating an overall visual effect, apart from the moral or narrative content of the painting. See Puttfarken, *Roger de Piles' Theory of Art.*
15. Puttfarken, *The Discovery of Pictorial Composition*, 226–227.
16. Puttfarken, it should be noted, sees these developments in French painting at the end of the eighteenth century somewhat differently. He emphasizes Jacques-Louis David's rebellion against an academic tradition of composition that by the end of the eighteenth century had become ever more restrictive—that, for example, the protagonist *always* had to be shown in the strongest light, that unity of time and place needed to be strictly observed, etc. See Puttfarken, "David's *Brutus* and Theories of Pictorial Unity in France."
17. Fried has argued that it was in the period we call "Romanticism" that the seeming inescapability of theatricality had ceased to be experienced as a merely artistic dilemma and had become instead a metaphysical threat to the very nature of humanity. "But unlike David then or later," Fried writes, "Géricault appears to have sensed in the theatrical a metaphysical threat not only to his art but also to his humanity (this was his romanticism)," in *Courbet's Realism* by Michael Fried (Chicago: University of Chicago Press, 1990), 23. See also Robert Pippin, "Authenticity in Painting," *Critical Inquiry* 31 no. 3 (Spring 2005), 575–598. For Pippin, the antitheatrical suspicion that figures have become overly posed returns to the Kantian problem of self-legislation that so preoccupied German Idealism. Kant was especially concerned with the question of how one could come to identify with moral law and not merely experience it as something imposed on us from outside.
18. See the sections "Leone Battista Alberti on Composition" and "The Human Body in Renaissance Criticism" in Puttfarken, *The Discovery of Pictorial Composition*, 53–62; and the chapter "Von der 'Komposition der Körper' zur Bildkomposition" in Hans Körner, *Auf der Suche nach der 'wahren Einheit,'* 12–71.
19. Puttfarken, *The Discovery of Pictorial Composition*, 56. See also Körner, *Auf der Suche*, 15–16.
20. Cited in Puttfarken, *The Discovery of Pictorial Composition*, 102 (my emphasis).
21. Shaftesbury recognized, in the English version of his article, that "picture" was an inadequate translation for *tableau*, instead coming up with the term "tablature." And he insisted that the *tableau* was distinct from "all those wider sorts of Painting" painted in fresco on the walls and ceilings of churches.

22. Shaftesbury, *A Notion of the Historical Draught or Tabulature of the Judgment of Hercules, According to Prodicus* (1713), 4. See Fried's discussion of this quote from Shaftesbury, and the relation of the *tableau* to organic unity, in *Absorption and Theatricality*, 89.
23. Linda Nochlin, *The Body in Pieces: Fragment as Metaphor* (London: Thames & Hudson, 2001), 19.
24. Nochlin, *The Body in Pieces*, 22.
25. Germer has noted that, according to Géricault's biographer Clément, the male head, which appears so alive, was actually that of a corpse, whereas the female head, which appears to be sleeping, was that of a live model: Germer, "Pleasurable Fear," 163.
26. Lennard Davis, "Nude Venuses, Medusa's Body, and Phantom Limbs," in *The Body and Physical Difference*, ed. David Mitchell and Sharon Snyder (Ann Arbor: University of Michigan Press, 1997), 51–70, 53–54.
27. James Porter, "Foreword" in *The Body and Physical Difference*, ed. David Mitchell and Sharon Snyder, xiii–xiv, xiii.
28. Nochlin has written compellingly about the ways this is possible in *The Body in Pieces*, 53–55.
29. Tobin Siebers, *Disability Aesthetics* (Ann Arbor: University of Michigan Press, 2010), 5.
30. Michael Fried, *Courbet's Realism*, 23.
31. Michael Fried, "Géricault's Romanticism," 647.
32. Fried writes in *Courbet's Realism*: "[T]he representation of animals, whether active or in repose, provided Géricault with something like a natural refuge from the theatrical, as if for him *the relation of animals to their bodies and to their world precluded* the theatricalizing of that relation no matter what," 24 (my emphasis).
33. See, for example, Lorenz Eitner's discussion of the reception of the *Medusa* and the personal crisis that followed in *Géricault: His Life and Work* (Ithaca, NY: Cornell University Press, 1983), 185–202.
34. According to Clément, the background was added not by Géricault but by his friend Horace Vernet. The scene is apparently based on a confrontation described in the liberal newspaper, *Le Constitutionnel*, in 1817. See Clément, 378.
35. Heidegger argued in *Being and Time* that tools only became visible as tools when they fell into disrepair. In this moment, they became obtrusive, troublesome. Contrast this, however, with his description of van Gogh's peasant shoes in "The Origin of the Work of Art" for a description of how the nature of tools might be represented in art.
36. Ernst Jentsch, "On the Psychology of the Uncanny," trans. Roy Sellars, *Angelaki* 2, no. 1 (1995), 7–16, 14.
37. Bruno Chenique, *La folie d'un monde* (Paris: Éditions Hazan, 2006).
38. Germer, "Pleasurable Fear," 165.
39. Heinrich von Kleist, "Über das Marionettentheater," in *Sämtliche Erzählungen* (Frankfurt am Main: Deutscher Klassiker Verlag, 2005), 555–563, 563. I have provided my own translation to emphasize specific themes.
40. Clément mistakenly identified the drawing as an illustration of a scene from *Macbeth* and described the woman as a "sorceress" whose curse is mocked by the two young people. See Clément, *Géricault*, 365.
41. Clément describes the peasant as "moins homme que bête" (less a man than an animal). To Clément, he appears brutish and inebriated (217).
42. In his classic biography of Géricault from the nineteenth century, Clément is at a loss for how to refer to this contraption. He calls it "une sorte de fauteuil grossier, de brouette à roues pleines" (a kind of crude chair, a *wheelbarrow* on solid wheels—my emphasis). See Clément, 217.

43. One is reminded of Géricault's lithograph of *Mazeppa* (1823), a heroic Romantic figure from Byron who was stripped naked, tied on his back to a horse, and released onto the steppe.
44. Nochlin has suggested that the older woman's paralysis may be the result of syphilis and thus directly related to sexual excess—this intimation of sexuality is further signified by her being strapped to a male figure. Nochlin, "Géricault, or the Absence of Women," *October* 68 (Spring 1994), 45–59, 56.
45. Eitner, *Géricault: His Life and Work*, 229. Contrast, however, my reading of the carefully composed structure of this lithograph with Eitner's claim of the "harsh incoherence of this scene, its lack of artistic arrangement."
46. I am grateful to the many interlocutors and readers who helped bring this essay about: First and foremost, to Ralph Ubl (University of Basel), without whom I might never have begun writing about painting and in conversation with whom I developed many of the ideas in this paper; to Robert Pippin and Jim Conant (University of Chicago), in whose graduate seminar I first began reading the work of Michael Fried; and to Hannah Eldridge (University of Wisconsin), for her careful and critical comments on the manuscript.

Part Two

FIGURES

4

ROMANTIC TALES OF PSEUDO-AUTOMATA

Jacques de Vaucanson and the Chess-Playing Turk in Literature and Culture

Wendy C. Nielsen

BY BRINGING THE WORKS OF GERMAN, French, and American literary and cultural history together, this essay shares the comparative focus of the contributions to *Romantic Automata: Exhibitions, Figures, and Organisms* and offers another study of automata in the work of E.T.A. Hoffmann that ties Romantic-era texts to modern criticism (Walter Benjamin and Karl Marx). It explores the metafictional aspects of automata in order to make the case that nonhuman figures provoke not only paranoia but also provide inspiration about the future of humans and their relation to work. Pseudo-automata suggest that exhibitions of such figures generate reflection on the extent to which labor turns the human organism into a type of machine.

The automaton is from its very definition *pseudo*, meaning fake or false, for it mimics the movement of humans and animals. As Dennis Des Chene explains: "Automaton was sometimes used in its root sense to denote anything capable of self-movement. In that sense, the soul—and God—are automata but not machines."[1] Recent monographs by cultural historians have expanded knowledge about the importance of eighteenth- and nineteenth-century automata to understanding some of the debates in that period around natural philosophy, human anatomy and physiology, and the aesthetics of musicianship and entertainment.[2] This essay examines the fictional nature of Romantic-era automata and interrogates the ways in which tales about ostensible androids, such as the Chess-Playing Turk, shape the discourse around the future of work and workers across the globe. The Chess-Playing Turk was a cultural phenomenon in the eighteenth and nineteenth centuries but was, in fact, operated by a human. German scholar Rebecca Wolf calls the Chess Player a pseudo-automaton [Pseudoautomat], a term I adopt for this essay.[3]

A literary analysis of automata, pseudo-automata, and their makers can reveal the various fictions involved in creating them. At least one source points to the importance of automata in the global market, although evidence is lacking for the claim that they play a pivotal role in international trade.[4] The wealth of the pseudo-automaton lies in its ability to generate ideas about ways to change the modern work environment through automation. Tales about pseudo-automata function as meta-narratives about the future of global commerce in Romantic-era Britain and Europe. The first half of this essay explores narratives about the French inventor Jacques de Vaucanson (1709–1782) in order to establish a common theme in tales about pseudo-automata: the inventor as an isolated progenitor who aims to present and profit from realistic performances of his creations. The second part of the essay explores tales about the Chess-Playing Turk, and a third section investigates the ways in which literature promotes skepticism about such creations. In texts written by Ernst Theodor Amadeus Hoffmann (1776–1822), Edgar Allan Poe (1809–1849), and Walter Benjamin (1892–1940), pseudo-automata represent writers' concerns about technology, work, and the unsuspecting public.

Tales of pseudo-automata predate the Romantic period. The most famous automaton of the early modern period may not, in fact, have even existed. The philosopher René Descartes (1596–1650) ostensibly created a replica of his deceased daughter, Francine. He reportedly brought her aboard a ship, but the captain, fearing its uncanny movements, threw it overboard. Historian Minsoo Kang finds this tale of Descartes's mechanical daughter in "no less than thirty-nine English and French works," including recent scholarly publications.[5] However, Descartes never constructed this automaton. Kang argues this "intellectual fable"[6] exists owing to the efforts of a Carthusian monk, Bonaventure d'Argonne (1634–1704), to obfuscate the scandal associated with Descartes having an illegitimate child.[7] A Romantic-era publication revived interest in Descartes's pseudo-automaton. Isaac Disraeli (1766–1848), father of the future Prime Minister Benjamin Disraeli, published the monk's tale in the first volume of his five-volume work, *Curiosities of Literature* (1791–1823), which became a common source of the story in English.[8] Kang surmises that the myth of Descartes's mechanical daughter persists today because it seems to substantiate current-day thinking of the philosopher "as a kind of proto-cybernetic thinker who anticipated many of today's questions" about artificial life.[9]

JACQUES DE VAUCANSON

The pursuit of artificial life in the twenty-first century aims to bring the apocryphal story of Descartes' daughter to life: to create an engineered humanoid that can simulate people. This goal can be traced to the early eighteenth century when automaton manufacture advances well beyond the garden automata and animated clocks of the early modern period and adopts a "simulative" turn towards design, as Jessica Riskin notes.[10] Yet while many inventions aimed to mimic human and animal anatomy, they stopped short of becoming exact models of live creatures. For example, the digesting duck [canard digérateur] of Jacques de Vaucanson appeared to eat corn and expel pellets. However, the duck only *appeared* to digest food, and the pellets were already present in the duck, ready for discharge.

Scholars seem to agree on the *intent* of Vaucanson's work to contribute to furthering "an understanding of nature's law (and God's existence)," as Joan Landes writes.[11] The duck intervened, according to Riskin, in debates "among physiologists over whether digestion was a chemical or a mechanical process," and included "over four hundred articulated pieces, imitating every bump on every bone of a natural wing."[12] Vaucanson had attended courses in anatomy and medicine,[13] and reportedly met with a surgeon, Claude-Nicolas Le Cat (1700–1768), who helped to inspire his work.[14] Vaucanson's biographers describe his plan to construct an automaton that models the biological functions of the human body, including "the circulation of blood, respiration, the play of muscles, tendons, nerves, etc."[15] The surgeon Le Cat had also expressed plans to create an artificial man that could model anatomy and speak.[16] Vaucanson's creations thus embodied the ostensible purposiveness of anatomical design.

The narrative of Vaucanson's duck emphasizes anatomical authenticity. In the widely circulated and translated brochure about his automata, *The Mechanism of the Automaton-Flute-Player: presented to the Gentlemen of the Royal Academy of Sciences* [*Le mécanisme du fluteur automate, presenté à Messieurs de l'Académie royale des sciences,* 1738], Vaucanson describes his attempt to imitate the natural processes of nature in the appended "Letter to the Abbe [*sic*] De Fontaine":

> The Duck stretches out its Neck to take Corn out of your Hand; it swallows it, digests it, and discharges it digested by the usual Passage. You see all the Actions of a Duck that swallows greedily, and doubles the Swiftness in this Motion of its Neck and Throat or Gullet to drive the Food into its Stomach, copied from Nature: The Food is digested as in real Animals, by Dissolution, not Trituration, as some natural Philosophers will have it.[17]

> [Il allonge son cou pour aller prendre du grain dans la main, il l'avale, le digere, & le rend par les voyes ordinaires tout digeré; tous les gestes d'un Canard qui avale avec précipitation, & qui redouble de vitesse dans le movement de son gosier, pour faire passer son manger jusque dans l'estomac, y sont copiés d'aprés nature: l'aliment y est digeré comme dans les vrais animeaux, par dissolution, & non par trituration, comme le prétendent plusieurs Physiciens].[18]

The translator, John T. Desgauliers, in the words of one scholar, "the doyen of London engineers and demonstrators,"[19] and, the title page announces, the "chaplain to his Royal Highness the Prince of Wales," conveys an intimate tone by translating Vaucanson's prose into a second-person narration ("You see all the Actions"). The epistolary address of this appended text also suggests a shared secret. This genuine tone might have facilitated readers' ability to connect with the sincerity of Vaucanson's purpose. David Brewster's *Letters on Natural Magic*, for example, reiterates Vaucanson's description of the duck's "process of digestion" almost verbatim.[20] Eighteenth-century tales about the digesting duck obscured its mechanical illusions because of observers like Brewster, who believed the duck actually consumed and digested corn.

Modern scholarship on automata focuses on the authenticity of their inner workings. For the scholar Paul Metzner, the lack of an actual "organic connection"[21] between the parts of eighteenth-century automata makes them what he calls quasi-automata. For many eighteenth-century observers, readers, and even inventors, the degree of realism offered in the *presentation* of automata seemed more important than the apparent authenticity of their inner construction. *The Memoirs of the Duke of Luynes*, for example, recalls disappointment that the mouth of Vaucanson's flute player remains permanently open, thus destroying the illusion of a true musical performance.[22] Audiences expected inventors to endeavor to depict physiology in a realistic fashion, even if the inner workings fell short. The *Encyclopedia* entry for *automaton* makes special note of Vaucanson's intent, although the author, co-editor Jean-Baptiste le Rond d'Alembert (1717–1783), remains ignorant about the fake excrement: "The inventor does not pretend that this digestion is perfect, capable of producing blood and nutritional fluids to sustain the animal, and it would be churlish to criticize him for it"[23] [L'Auteur ne donne pas cette digestion pour une digestion parfaite, capable de faire du sang & des sucs nourriciers pour l'entretien de l'animal; on auroit mauvaise grace de lui faire ce reproche].[24] Vaucanson's faithful representation of the operation of digestion holds greater importance for writers like d'Alembert than his ability to recreate digestion accurately.

On the other hand, the generation after the Enlightenment recognized that convincing performances threatened to trick less than savvy observers. *The Automaton of Vaucanson* [*L'automate de Vaucanson*], a comic opera in one act by Adolphe de Leuven (1802–1884), premiered at l'Opéra-Comique in Paris on September 2, 1840. It dramatizes the effect of a pseudo-automaton on a woman and a servant. The main action of the play concerns Vaucanson's niece, Marie, whom Chevalier de Lancy woos by impersonating Vaucanson's automaton. Even Vaucanson's worker, Landry, cannot distinguish between this impersonation of an automaton and the machine, and he believes the masses' accusation of Vaucanson: that the automaton is possessed, and Vaucanson is a sorcerer. In the final scene, Chevalier de Lancy returns, and Marie and Landry recognize his voice, revealing that the automaton is, in fact, "a fake automaton" [un automate postiche].[25] Some of the advances in sound technology in the late eighteenth and early nineteenth century make the premise of Leuven's comic opera seem probable. Erasmus Darwin (1731–1802) built mechanical mouths with "leather lips, valves for nostrils, and a ribbon for a tongue."[26] His friend and fellow member of the Lunar Society, Richard Lovell Edgeworth (1744–1817), recalls in a letter to Darwin: "I placed one your mouths in a room near some people in 1770, who actually thought I had a child with me, calling papa and mamma."[27] *The Automaton of Vaucanson* plays on the potential of the cleverly exhibited automata to dupe unsuspecting observers and spotlights the depiction of the inventor as a kind of genius, which occurs in other French accounts of Vaucanson.[28]

Vaucanson's career demonstrates that while automata begin circulating in the eighteenth century as trivial luxury items, they ultimately affect the future of how humans work. Four years after the first exhibit in 1738, Vaucanson sold his automata to a consortium of investors in Lyon,[29] who exhibited them in London and across Europe;[30] the automata eventually separated and fell into various collectors' hands.[31] By 1749, Vaucanson presented a plan for an automatic loom to the Academy of Sciences and became the national inspector of silk manufacture.[32] Riskin interprets Vaucanson's new position as the government's attempt to preserve the inventor's role as a *French* asset, since it drew him away from accepting a lucrative position at the court of Frederick the Great for "a generous salary of 12,000 livres."[33] (Reed Benhamou estimates the annual earnings of the average laborer in the eighteenth century at under 160 livres.)[34]

The silk workers of Lyon, however, disliked the plan to replace their work with automatic spinning looms. They rioted and forced Vaucanson to flee the city disguised as a monk.[35] While this riot predated the Luddite rebellions in England by several decades, machine-breaking was a long-standing tactic that became more and more commonplace in the latter half of the eighteenth century.[36] Writing and literature articulate the changing narrative of work in which the plot concerns not merely man *versus* machine but man *as* a machine. As the Scottish philosopher and historian Adam Ferguson (1723–1816) writes in *An Essay on the History of Civil Society* (1767): "Manufactures, accordingly, prosper most, where the mind is least consulted, and where the workshop may, without any great effort of imagination, be considered as an engine, the parts of which are men."[37] In the narrative of modern work, automated machines threatened not just to make workers obsolete, but also to transform workers into machines. The Chess-Playing Turk further blurred the lines between man and machine by suggesting the possibility that machines could think like humans.

THE CHESS-PLAYING TURK

Wolfgang von Kempelen (1734–1804), a native Hungarian from Pozsony (Bratislava), created the Chess-Playing Turk in the spring of 1770 at the court of Maria Theresa in Vienna, where as councillor he translated the Hungarian civil code into German, founded a fabric printing workshop, directed the state-owned salt mines, constructed water pumps for castles, presided over the opening of a theater in Buda, and wrote at least one drama.[38] Kempelen had previously constructed and written about a Speaking Machine that could produce around 30 words and a few phrases.[39] Like Vaucanson, Kempelen studied anatomy in order to fashion it as authentically as possible.[40] The writer Baron Friedrich Melchior Grimm (1723–1807) recalls its voice as "pleasant and sweet" [agréable et douce].[41] However, Johann Wolfgang von Goethe (1749–1832) found the machine "not especially eloquent" [zwar nicht sehr beredt]; he noted its ability to produce "various childish words and tones quite dutifully" [verschiedne kindische Worte und Töne ganz artig].[42] Goethe's term for dutiful, *artig*, is a word often used to describe children and typifies what Kara Reilly identifies as the eighteenth-century attitude towards automata as "ideal children."[43] In fact, Kempelen's Speaking Machine (now in the Deutsches Museum in Munich) evoked childishness only in its tone of voice but stopped short of mimicking a child's form, although according to his friend and promoter Karl Gottlieb von Windisch (1725–1793), he intended to make the machine resemble a five- or six-year-old child.[44]

Kempelen's most famous automaton, the Chess-Playing Turk, appeared to be so advanced that contemporaries considered it among the first humanoid automata, or androids.[45] The Chess Player wore a Westernized stereotype of Turkish attire, smoked a pipe, and sat behind the chessboard atop a chest that contained clockwork. Kempelen constructed the box to be opened to show that no one was inside and wound a casket every few moves in a way that suggested that he was operating a machine. In 1783, Kempelen made the Chess Player portable and took it on tour through parts of German-speaking Europe, Paris, and London. Notable personages played against the Chess Player, including Benjamin Franklin and Napoleon.[46] After Kempelen's death, Napoleon's stepson, Eugène Beauharnais, bought the Chess Player and then sold it to Johann Nepomuk Maelzel (1772–1838), who toured the Chess Player in America along with other genuine automata and entertainments.

For writers, the Chess Player evoked anxiety about the potential of machines to replace humans in various activities, especially labor, and raised the issue of the way in which most machines differed from humans: the ability to think. The German author Jean Paul (Johann Paul Friedrich Richter, 1763–1825) juxtaposes these ideas in his satirical piece from 1789, "Our Humble Idea about All the Players and Speaking Ladies of Europe, Contrary to and against the Introduction of Kempelen's Playing and Speaking Machines" [Unterthänigste Vorstellung unserer, der sämmtlichen Spieler und redenden Damen in Europa, entgegen und wider die Einführung der Kempelischen Spiel- und Sprachmaschinen], in *Selection from the Devil's Papers* [*Auswahl aus des Teufels Papieren*]:

> if only he [Kempelen], instead of sitting down and fervently hatching speaking and playing machines, which at a single blow take food away from thousands of his brothers, had truly contemplated and brought about thinking machines; for since only very few professions relate to thinking, then he would have been able to wreak very little or no havoc, since particularly the few that would appear to starve through rivalry with thinking machines, surely would have died without this hunger.
>
> [wenn er, anstatt sich niederzusetzen und feurig Sprach- und Spielmaschinen auszubrüten, die auf einmal tausend seiner Brüder außer Nahrung setzen, recht nachgesonnen hätte und wirklich mit Denkmaschinen zum Vorschein gekommen wäre; denn da nur sehr Wenige Profession vom Denken machen, so hätt' er geringes oder kein Unheil anrichten können, da zumal die Wenigen, die durch die Nebenbuhlerei der Denkmaschinen verhungert zu sein geschienen hätten, sicher auch ohne diese Hungers gestorben wären.][47]

To those who believed in its ruse, the Chess Player represented an evolutionary step in machines, that is, the ability to think. Of course, Jean Paul's satire reminds readers of this crucial difference between man and machine. Yet the text also conveys the real concern that machines have the potential to take work away from human beings through automation. Indeed, the Chess Player apparently inspired the automation of weaving. The English inventor Edmund Cartwright (1743–1823), according to his memoir, thought that if a "most complicated machine" like the Chess Player could exist, then so too could steam-powered looms, which he eventually produced for factories in Manchester.[48] Kempelen had similar ambitions; in London, he sought advice on his steam engine from James Watt's partner, Matthew Boulton.[49] So while the automaton represents, according to Catherine Liu, a "nonproductive machine," it still "inspires both automation and mass production."[50]

Tales about automata, even pseudo-automata like the Chess Player, have the power to change the way humans work and interact with machines and each other.

On the other hand, Jean Paul's remarks also seemed to undermine the Chess Player's status as a machine that could think. Many other observers had long suspected that the Chess-Playing Turk was not what it seemed: one of the first androids able to make reasoned decisions.[51] In 1784, a French author, Henri Decremps (1746–1826), first suggested that a dwarf was hidden inside.[52] The fact that the Chess Player *lost* so often led to suspicion, however; indeed, there was a full-grown operator who sat concealed behind a second panel. The clockwork mechanisms and hand-held casket, while lending the automaton an air of mechanical complexity, played no real function in its operation. Several other authors wrote about the Chess Player and speculated on how it worked,[53] and even the theater played a role in the legend of the Chess Player. A comedy by the German actor Heinrich Christian Beck, *The Chess Machine* [*Die Schachmaschine*, 1798], nearly guesses the secret of the Chess Player; a character hides inside it and reveals himself in the final act.[54]

The Chess Player is thus an *inverse* of Olimpia in E.T.A. Hoffmann's novella, "The Sandman" ["Der Sandmann," 1817]. Whereas Olimpia is a machine pretending to be human, the Chess Player is a human pretending to be a machine. Problems with human anatomy sometimes made the Chess Player's passing as a machine difficult. The work took place in dark, cramped, and uncomfortably warm conditions. The operator had a candle, and its smoke went up through the Chess Player's pipe. The operator could thus follow the play by looking up through the body of the Chess Player, and magnets helped to identify the location of pieces on the chessboard. The Chess Player also operated as a pseudonym for a series of operators. One of the Player's operators was the Parisian chess player Hyacinthe Henri Boncourt (d. 1840), who, at six feet, could not operate the machine for long periods owing to the cramped conditions;[55] Boncourt struggled not to sneeze, and so Maelzel installed a "noisy spring."[56]

Interestingly, the Chess Player was also an android in drag. At least one woman operated the Chess Player. Her name is unknown, but one source claimed she was the wife of the Frenchman in charge of Maelzel's ropedancers.[57] Since she was a novice at chess, Maelzel taught her to play endgames before their ship arrived in New York. Playing only endgames was convenient for the operator because it shortened the length of play and made working in the tiny, hot space of the Chess Player's cabinet less of a challenge. Nearly all of the Chess Player's operators kept its secrets except Jacques-François Mouret (1787–1837), great-nephew of the famous chess player, Philidor, and Maelzel's operator in London in 1819. The alcoholic Mouret apparently sold his story to *Le Magasin pittoresque* in 1834 and died penniless three years later.[58]

While automata help bolster the careers of inventors like Vaucanson and Kempelen, it is not clear to what extent they actually lead to the private accumulation of capital in the case of Maelzel. According to Paul Metzner's *Crescendo of the Virtuoso: Spectacle, Skill, and Self-Promotion in Paris during the Age of Revolution* (1998), "Maelzel died a millionaire at sixty-six known to posterity as the inventor of the metronome."[59] However, Metzner provides no proof for this assertion, and further clarification is needed. Maelzel patented the metronome after allegedly stealing the idea from its inventor, Dietrich Nikolaus Winkel (1777–1826), who later sued him.[60] Maelzel was apparently in debt to the previous owner of the automaton, Beauharnais, who had also sued Maelzel for not making payments, and his flight to America in 1826 was likely motivated at least in part to escape the pressure of this debt, as well as Winkel's suit against him.[61] Maelzel pirated another person's work on at least one other occasion. He designed ear trumpets for Beethoven, and in 1813, Beethoven wrote *Wellington's Victory* for Maelzel's panharmonicon. Predictably, their friendship fell apart when Maelzel stopped crediting Beethoven for his piece.[62] The Chess Player can be read as an extension of Maelzel's habit of plagiarizing; as James Berkley notes, "an automaton already is a plagiarism: definitionally, it is a mechanical imitation of life."[63]

In 1837, Maelzel went on tour in Havana, Cuba after the truth about the Chess Player started circulating in American newspapers. One of these sources is Edgar Allan Poe's piece for the *Southern Literary Messenger*, "Maelzel's Chess Player" (1836), which suggests the identity of one of the operators: Maelzel's secretary, William Schlumberger.[64] The truth was destined to remain obscure, since Schlumberger died of yellow fever in Cuba, and Maelzel died on the return voyage.[65] Thereafter, the Chess Player and other automata were auctioned, and one of Maelzel's creditors purchased it; he in turn sold it to a consortium of Philadelphians who wanted to discover its secret; later, they donated it to the Chinese Museum in Philadelphia, where a fire destroyed it in 1854.[66] The Chess-Playing Turk lived on in cultural memory, however, through literary representations and at least two films.[67]

HOFFMANN, POE, AND BENJAMIN

In literature, the Chess Player sometimes serves as a way to frame questions about technology and culture through the lens of irony; it can symbolize the potential of cleverly displayed automata to confuse vulnerable audiences and to make human workers obsolete and/or more like machines. The Chess Player appears as the Talking Turk in Hoffmann's short story, "Automata" ["Die Automate"]. Originally published in the *Allgemeine Musikalische Zeitung* in 1814, the text was later incorporated as a fragmentary narrative into Hoffmann's collection of tales, *The Serapion Brethren* [*Die Serapionsbrüder*, 4 vols., 1819–1821]. In "Automata," Professor X's automaton, the Talking Turk, appears to possess the ability to predict the future. One of the two main characters, Ferdinand, believes the Turk's prediction that when he sees his beloved again, she will be lost to him; he leaves the company of his friend, Ludwig, and becomes entranced when he hears a woman singing, whom he later sees in the company of Professor X. It is not clear if the unnamed woman is yet another automaton, and if Ferdinand, like Nathanael in "The Sandman," is succumbing to agalmatophilia, or sexual attraction to dolls.

By giving the Turk the power of speech, Hoffmann anticipates one of Maelzel's modifications to the automaton; between 1819 and 1820, the Chess Player started saying "check" or "*échecs*" in French.[68] Hoffmann's source, Johann Christian Wiegleb's *Natural Magic* [*Die natürliche Magie*, 1789], claimed that the Chess Player could audibly answer questions.[69] Another German text, Johann André's musical comedy, *The Automaton* [*Das Automat*, 1787], makes a similar assertion: that the automaton acts as an oracle.[70]

The bifocal view that typifies Romantic irony[71] and characterizes Hoffmann's "The Sandman"[72] is already present in "Automata." Irony in "Automata" serves to caution readers about trusting too much in the illusion that automata present. Whereas Ferdinand is duped by Professor X's handiwork, his friend, Ludwig, remains skeptical, just as in "The Sandman" the perspective of the editor, and Nathanael's friends, Lothar and Siegmund, undermine Nathanael's naïveté. In "Automata," Ludwig rails against the lack of sensibility in musical automata and cites Vaucanson's flute player to underscore his sense that "all mechanical music seems monstrous and abominable to me"[73] [vollends [ist] die Maschinenmusik . . . für mich etwas heilloses und gräuliches].[74] Yet, major German composers wrote pieces for musical automata in the eighteenth and nineteenth centuries.[75] Hoffmann critiques mechanical music through two narratives: through Ludwig's argument and by suggesting that mechanical music is as "monstrous and abominable" as a young man seeking love from an illusion. The narrative style directs the reader to sympathize more with Ludwig's skepticism about automating musicianship rather than Ferdinand's credulity because the latter disappears to search for his beloved, who travels in the company of Professor X.

The German verb "türken," literally "to turk," means to fabricate or falsify, and at least one source attributes this connotation to Kempelen and Maelzel's Chess-Playing Turk.[76] Perhaps this is why the creator of the Talking Turk in "Automata," Professor X, has a "mountebank manner" [marktschreierische Art].[77] What business might professors have in creating automata that mimic human behavior? Dieter Müller argues that the figure of the professor in Hoffmann represents the "mechanical-automatic Enlightenment worldview, that ultimately works destructively,"[78] but professors in Hoffmann can also be read as liminal figures in protoindustrial society. In "The Sandman," Spalanzani, a professor of physics modeled on the scientist Lazzaro Spallanzani (1729–1799), dupes Nathanael into believing in Olympia's personhood, and in a letter to Klara, Nathanael likens him to the greatest huckster of the eighteenth century, Alessandro Cagliostro (1743–1795).[79]

Perhaps Professor X and Professor Spalanzani resemble hucksters for good reason: economic need. Historian Charles McClelland notes while the number of full professors increased 35 percent in Prussian universities between 1775 and 1804,[80] the professoriate lived on low salaries and remained somewhat isolated from mainstream society; professors had a reputation for "quarrelsomeness" and tended to dress differently, which made them appear "peculiar."[81] Many universities in what is now Germany compensated professors "with goods and privileges" such as "firewood, food products . . . or licenses to the professors to maintain beer and wine cellars for public sales."[82] In such an atmosphere, research projects likely became important factors in maintaining professors' financial stability. The ability to make ends meet was something the peripatetic Hoffmann might well have sympathized with as he moved between a series of jobs in the Prussian courts and in Bamberg's musical theater. The words of a professor of Poesy and Eloquence in "The Sandman" could perhaps best describe Hoffmann's image of professors: "The whole thing is an allegory—an extended metaphor!—You understand me!—A word to the wise is enough!" [Das Ganze ist eine Allegorie—eine fortgeführte Metapher!—Sie verstehen mich!—*Sapienti sat*!].[83]

My reading of Hoffmann's relation to technology thus differs from that of the philosopher Walter Benjamin. In his essay, "E.T.A. Hoffmann and Oskar Panizza" (1930), Benjamin describes Hoffmann as the "pedantic, proper court councillor" who equates "the satanic" with "the mechanical."[84] Perhaps Benjamin projects some of his own preoccupations about capitalism onto Hoffmann. He begins his "On the Concept of History" ["Über den Begriff der Geschichte," 1940] with an analogy of the Chess-Playing Turk that makes technology seem insidious; he associates the hidden operations of the Chess Player with the ways in which workers help obfuscate the forces of capitalism:

> There was once, we know, an automaton constructed in such a way that it could respond to every move by a chess player with a countermove that would ensure the winning of the game. A puppet wearing Turkish attire and with a hookah in its mouth sat before a chessboard placed on a large table. A system of mirrors created the illusion that this table was transparent from all sides. Actually, a hunchbacked dwarf—a master at chess—sat inside and guided the puppet's hand by means of strings. One can

> imagine a philosophic counterpart to this device. The puppet, called "historical materialism," is to win all the time. It can easily be a match for anyone if it enlists the services of theology, which today, as we know, is wizened and has to keep out of sight.[85]

> [Bekanntlich soll es einen Automaten gegeben haben, der so konstruiert gewesen sei, daß er jeden Zug eines Schachspielers mit einem Gegenzuge erwidert habe, der ihm den Gewinn der Partie sicherte. Eine Puppe in türkischer Tracht, eine Wasserpfeife im Munde, saß vor dem Brett, das auf einem geräumigen Tisch aufruhte. Durch ein System von Spiegeln wurde die Illusion erweckt, dieser Tisch sei von allen Seiten durchsichtig. In Wahrheit saß ein buckliger Zwerg darin, der ein Meister im Schachspiel war und die Hand der Puppe an Schnüren lenkte. Zu dieser Apparatur kann man sich ein Gegenstück in der Philosophie vorstellen. Gewinnen soll immer die Puppe, die man "historischen Materialismus" nennt. Sie kann es ohne weiteres mit jedem aufnehmen, wenn sie die Theologie in ihren Dienst nimmt, die heute bekanntlich klein und häßlich ist und sich ohnehin nicht darf blicken lassen.][86]

The editors of *Walter Benjamin: Selected Writings* (2006) attribute the above translation to Harry Zohn. Zohn died in 2001, five years before the book's publication. In his 1955 translation of this same text, Zohn translates "hunchbacked dwarf" [buckliger Zwerg] as "little hunchback," leaving out the word *dwarf*.[87] However, recent scholarship stresses the importance of the figure of the hunchbacked dwarf in Benjamin's writing.[88] In fact, it is Benjamin who inserts the hunchbacked dwarf into the narrative of the Chess Player, for the figure is dismissed in his source: Charles Baudelaire's translation of Poe's essay, "Maelzel's Chess Player," as "Le joueur d'échecs de Maelzel" (1856). Poe already discounted the hypothesis that a dwarf had operated the machine, calling the idea "too obviously absurd to require comment."[89]

Yet in Benjamin's recounting, the ostensible dwarf operator takes on a sinister aspect with its "hunchbacked" [buckliger] form. Robert Gold surmises that for Benjamin, "the little hunchback signifies forgetting, his deformed shape embodying the distortion imposed upon all that is irretrievably lost to the past,"[90] particularly "the ability of the proletariat to hear echoes of its own struggle in the past, a prerequisite for revolution."[91] Michael Löwy reads the term *theology* as referring "to two fundamental concepts: remembrance (*Eingedenken*) and messianic redemption (*Erlösung*)."[92] What's at stake in winning the game here, according to Löwy, is "struggling against the oppressors' view of history" and "defeating the historic enemy itself, the ruling classes."[93] The hunchbacked dwarf ostensibly operating the Chess Player symbolizes the worker's complicity with the sinister forces of dialectical history, which in 1940 essentially equates with fascism. In this way, Benjamin's revision of the tale to spotlight the dwarf is deliberate; his metaphor illustrates that modern capitalism transforms the worker into a hidden person, or a pseudo-person.

In the twenty-first century, Benjamin's thesis seems to foreshadow the ways in which the apparatus of technology sometimes hides the contributions of human workers. The online shopping website, Amazon.com, offers a labor service called "the Mechanical Turk" that connects workers, colloquially called "Turkers," with employers seeking to complete "Human Information Tasks (HITs)" such as data processing.[94] Scholar Lilly Irani questions the extent to which the Amazon Mechanical Turk enables companies to appear technology-driven when they in fact rely on human workers.[95] Yet, in the context of our discussion, Amazon's name for its service is appropriate. Similarly, Maelzel and Kempelen's Chess-Playing Turk also hid the work of its human operators. Indeed, in nineteenth-century writing, the word "automaton" connoted work. In *Capital* [*Das Kapital*, 1867], Karl Marx and Friedrich Engels call the operation of weaving machines "a huge automaton" [einen großen Automaten].[96] For their understanding of the machine's role in mechanized cotton manufacture, Marx and Engels cite Andrew Ure's book, *The Cotton Manufacture of Great Britain* (1836), published the same year as Poe's "Maelzel's Chess Player," the text that informed Benjamin's first thesis.[97]

Poe saw the Chess Player perform in Richmond, Virginia in 1836, the same year he married his thirteen-year-old cousin, Virginia Clemm.[98] "Maelzel's Chess Player" aims to uncover the secrets of the Chess-Playing Turk. Poe's essay resembles creative nonfiction in its adherence to the facts; he analyzes previous writing on the Chess Player, which he refers to as "the Automaton," and his text employs ratiocination, a technique used in his detective fiction, as Klaus Benesch points out.[99] In Poe's narrative, Maelzel and Kempelen emerge as rather devious designers who dupe audiences by making the automaton's actions appear less than lifelike and therefore more machine-like: "Were the Automaton life-like in its motions, the spectator would be more apt to attribute its operations to their true cause, (that is, to human agency within) than he is now, when the awkward and rectangular manœuvres convey the idea of pure and unaided mechanism."[100] So while Poe recognizes the Chess Player as a pseudo-automaton, he attributes its ability to convince audiences of its authenticity to its mechanical mannerisms.

CONCLUSION

Poe's remark on the apparent lifelessness of the Chess Player suggests that after over sixty years of operation, three owners, and decades of touring Europe, Britain, and then America, the automaton was showing its age. Goethe complained about a similar aspect of Vaucanson's duck; by the time he saw it in 1805, it could consume but could no longer simulate digestion by expelling pellets.[101] Vaucanson's duck and the Chess Player were highly singular pieces, and only the continued spreading of tales about them guaranteed their circulation. While Liu is correct that automata became obsolete owing to their artisanal nature, the cultural memory afforded by literature reproduces them in perpetuity.[102]

Automata and pseudo-automata in the Romantic period served an important purpose: to provoke audiences to think about the ways that technology might reshape the future. The Chess-Playing Turk appeared to surpass audiences' expectations in that it presented a machine that could apparently make decisions. Authors like Hoffmann, Poe, and Benjamin remained skeptical about the value of believing in such fictions, however. Literature about pseudo-automata reveals that the convincing performances of Vaucanson's creations and the Chess-Playing Turk contain a maze of fictional clockwork that raises questions about the trustworthiness of the inventor-creator. Romantic tales about pseudo-automata suggest that automated work has the potential to turn humans into pseudo-people. In this way, Romantic tales of pseudo-automata critique the ways in which even fictitious automation has the potential to change work and the lives of workers across the globe.

NOTES

1. Dennis Des Chene, *Spirits and Clocks: Machine and Organism in Descartes* (Ithaca: Cornell University Press, 2001), 13.
2. See Stefani Engelstein, *Anxious Anatomy: The Conception of the Human Form in Literary and Naturalist Discourse* (Albany: SUNY Press, 2008); Minsoo Kang, *Sublime Dreams of Living Machines: The Automaton in the European Imagination* (Cambridge: Harvard University Press, 2011); Catherine Liu, *Copying Machines: Taking Notes for the Automaton* (Minneapolis: University of Minnesota Press, 2000); Paul Metzner, *Crescendo of the Virtuoso: Spectacle, Skill, and Self-Promotion in Paris during the Age of Revolution* (Berkeley: University of California Press, 1998); Kara Reilly, *Automata and Mimesis on the Stage of Theatre History* (New York: Palgrave Macmillan, 2011); Jessica Riskin, *The Restless Clock: A History of the Centuries-Long Argument Over What Makes Living Things Tick* (Chicago: University of Chicago Press, 2016); and Adelheid Voskuhl, *Androids in the Enlightenment: Mechanics, Artisans, and Cultures of the Self* (Chicago: University of Chicago Press, 2013); see also the contributions by Erin M. Goss and Frederick Burwick in this volume who focus on some of the same automata through the lens of feminist criticism and theories of the "uncanny," respectively.
3. Rebecca Wolf, *Friedrich Kaufmanns Trompeterautomat. Ein musikalisches Experiment um 1810* (Stuttgart: Franz Steiner, 2011), 16.
4. According to Richard D. Altick, James Cox's sale of automata to the Chinese court temporarily helped "reduce Britain's trade deficit" by offsetting its large orders of tea: *The Shows of London* (Cambridge: Harvard University Press, 1978), 69. However, Altick's source refers to an unneeded glut of English toys, known as "sing-songs," being sold in China:

Hosea Ballou Morse, *The Chronicles of the East India Company Trading to China, 1635–1834*, vol. 5 (Cambridge: Clarendon, 1929), 154.

5. Minsoo Kang, "The Mechanical Daughter of Rene Descartes: The Origin and History of an Intellectual Fable," *Modern Intellectual History* (19 Aug. 2016), 1–28, 8.
6. Kang, "The Mechanical Daughter of Rene Descartes," 4.
7. Kang, "The Mechanical Daughter of Rene Descartes," 10–11.
8. Kang, "The Mechanical Daughter of Rene Descartes," 15. The tale appears under the subtitle "Curious Automata" in Isaac Disraeli, *Curiosities of Literature*, 4th ed. (London: Murray and Highley, 1798), 507–510.
9. Kang, "The Mechanical Daughter of Rene Descartes," 8.
10. Jessica Riskin, "Eighteenth-Century Wetware," *Representations*, 83, no. 1 (Summer 2003): 97–125, 120.
11. Joan B. Landes, "'Tableaux quasi-vivants' and Simulacra. Anatomy and Design in Vaucanson's Automata," in *L'Automate, modèle, métaphore, machine, merveille* (Pessac: Presses Universitaires de Bordeaux, 2013), 189–206, 203.
12. Jessica Riskin, "The Defecating Duck, or, The Ambiguous Origins of Artificial Life," *Critical Inquiry* 29, no. 4 (Summer 2003): 599–633, 609.
13. Riskin, *The Restless Clock*, 118.
14. Joan B. Landes, "The Anatomy of Artificial Life: An Eighteenth-Century Perspective," in *Genesis Redux: Essays in the History and Philosophy of Artificial Life*, ed. Jessica Riskin (Chicago: University of Chicago Press, 2007), 96–118, 101.
15. "la *circulation du sang*, la respiration, *la digestion*, le jeu des muscles, tendons, nerfs, etc." André Doyon and Lucien Liaigre, *Jacques Vaucanson, Mécanicien de Génie* (Paris: Presses Universitaires de France, 1966), 148. Reed Benhamou claims an associate saw a model of a human in Vaucanson's studio in 1762 in his article, "From Curiosité to Utilité: the Automaton in Eighteenth-Century France," *Studies in Eighteenth-Century Culture* 17 (1987): 91–105, 102.
16. Riskin, "Eighteenth-Century Wetware," 114. Jonathan Sterne asserts that Le Cat actually built this "working model of the circulatory system" in *The Audible Past: Cultural Origins of Sound Reproduction* (Durham: Duke University Press, 2003), 73.
17. Jacques de Vaucanson, *An Account of the Mechanism of an Automaton, or Image Playing on the German-Flute*, trans. John T. Desaguliers (London: T. Parker, 1742), 21.
18. Jacques de Vaucanson, *Le mécanisme du fluteur automate, presenté à Messieurs de l'Académie royale des sciences* (Paris: J. Guerin, 1738), 19.
19. Simon Schaffer, "Enlightened Automata," in *The Sciences in Enlightened Europe*, ed. William Clark et al. (Chicago: University of Chicago Press, 1999), 126–165, 136.
20. David Brewster, *Letters on Natural Magic Addressed to Sir Walter Scott* (London: John Murray, 1842), 268.
21. Metzner, *Crescendo of the Virtuoso*, 177.
22. *Mémoires du Duc de Luynes sur la cour de Louis XV (1735–1758)*, ed. L. Dussieux et al., vol. 2 (Paris: Firmin Didot, 1860–65), 103.
23. Jean-Baptiste le Rond d'Alembert, "Automaton," *The Encyclopedia of Diderot & d'Alembert Collaborative Translation Project*, trans. Nelly S. Hoyt and Thomas Cassirer (Ann Arbor: Michigan Publishing, University of Michigan Library, 2003), accessed March 28, 2019, http://hdl.handle.net/2027/spo.did2222.0000.140.
24. Jean-Baptiste le Rond d'Alembert, "Automate," in *Encyclopédie, ou dictionnaire raisonné des sciences, des arts et des métiers, etc.*, ed. Denis Diderot et al., vol. 1, ARTFL Encyclopédie Project, ed. Robert Morrissey et al. (Spring 2016), 896–897, 896.
25. Adolphe de Leuven, *L'automate de Vaucanson* (Paris: Henriot et Cie, 1840), 9.
26. Sterne, *The Audible Past*, 74.

27. Richard Lovell Edgeworth, *Memoirs of Richard Lovell Edgeworth*, ed. Maria Edgeworth (London: Richard Bentley, 1844), 356.
28. Voltaire's sixth discourse "On the Nature of Man" ["Sur la nature de l'homme," 1738] compares Vaucanson to Prometheus.
29. Doyon and Liaigre, *Jacques Vaucanson*, 87.
30. Tom Standage, *The Turk: The Life and Times of the Famous Eighteenth-Century Chess-Playing Machine* (New York: Walker & Co., 2002), 12.
31. Friedrich Nicolai includes an account of the trade of Vaucanson's automata in *Beschreibung einer Reise durch Deutschland und die Schweiz, im Jahre 1781* (Berlin and Stettin: n.p., 1787), accessed March 28, 2019, https://babel.hathitrust.org/cgi/pt?id=nyp.33433066654074;view=1up;seq=7. Gottfried Christoph Beireis, a collector in Helmstedt, later acquired the duck; Voskuhl, *Androids in the Enlightenment*, 174.
32. Doyon and Liaigre, *Jacques Vaucanson*, 487.
33. Riskin, *The Restless Clock*, 146.
34. Benhamou, "From Curiosité to Utilité," 34.
35. Standage, *The Turk*, 12.
36. Brian Bailey makes note of the riots against engine looms in 1675, and attacks against John Kay's "flying shuttle" in Bury, Lancashire in 1753, and Hargreaves' "Jenny" in Blackburn in 1768 in his book, *The Luddite Rebellion* (New York: NYU Press, 1998), 11.
37. Adam Ferguson, *An Essay on the History of Civil Society* (Dublin: Boulter Grierson, 1767), 273.
38. Alice Reininger, *Wolfgang von Kempelen: A Biography*, trans. Peter Waugh (New York: Columbia University Press, 2012), 13, 15, 56, 72, 191, 197, 208, and 251.
39. Wolfgang von Kempelen, J. G. Mansfeld, and Heinrich Füger, *Wolfgangs von Kempelen K.k. Wirklichen Hofraths Mechanismus der Menschlichen Sprache: Nebst der Beschreibung seiner sprechenden Maschine; Mechanismus der Menschlichen Sprache* (Vienna: J. V. Degen, 1791) and B. A. Dupl, *The History and Analysis of the Supposed Automaton Chess Player of M. De Kempelen* (Boston: Hilliard, Gray and Co., 1826), 13.
40. Reininger, *Wolfgang von Kempelen*, 228.
41. See the September 1783 entry of Friedrich Melchior Grimm, *Correspondance littéraire, philosophique, et critique*, vol. 13 (Paris: Chez Furne, 1830), 449.
42. Goethe in a letter to Herzog Carl August on 12 June 1797, *Goethes Briefe*, vol. 2 (Hamburg: Christian Wegner, 1962), 277.
43. Reilly, *Automata and Mimesis on the Stage of Theatre History*, 91.
44. Karl Gottlieb von Windisch, *Briefe über den Schachspieler des Herrn von Kempelen* (Preßburg: A. Löwe, 1783), 28.
45. The first citation in the *OED* for "android" refers to an android made by Albertus Magnus, and the second citation refers to Kempelen's Chess Player. In the *Encyclopedia* entry for "Androide," d'Alembert and Diderot mention Vaucanson's flute player [*Flûteur automate*]; "Androide," *Encyclopédie, ou dictionnaire raisonné des sciences, des arts et des métiers, etc.*, ed. Denis Diderot et al., ARTFL Encyclopédie Project, ed. Robert Morrissey et al. (Spring 2016), 448–451, accessed March 28, 2019, https://encyclopedie.uchicago.edu/.
46. B. A. Dupl, *The History and Analysis*, 13.
47. Jean Paul, "Auswahl aus des Teufels Papieren," in *Jean Pauls Werke*, vol. 4 (Berlin: Gustav Hempel, 1870), 1–301, 65.
48. Mary Strickland Cartwright, *A Memoir of the Life, Writings, and Mechanical Inventions of Edmund Cartwright* (London: Saunders and Otley, 1843), 57. Cartwright was inspired in 1784, and the factory was constructed 15 years later; Standage, *The Turk*, 69–70.
49. Boulton refused to give Kempelen any tips about the steam engine and suggested his calculations were faulty; Reininger, *Wolfgang von Kempelen*, 215.

50. Liu, *Copying Machines*, x.
51. One of Pierre Jaquet-Droz's early creations, a pendulum clock with a "mechanical bell-striker" called "The Black Man," appeared, according to Voskuhl, to "'understand' and respond to inquiries about numbers worked by means of magnets": Voskuhl, *Androids in the Enlightenment*, 57.
52. Henri Decremps, *La magie blanche dévoilée ou explication des tours surprenants* (Paris: Langlois, 1784), 87 and in English, *The Conjurer Unmasked, Or, La Magie Blanche Dévoilée: Being a Clear and Full Explanation of All the Surprizing Performances Exhibited as Well in This Kingdom as on the Continent*, trans. T. Denton (London: C. Stalker, 1785), 95.
53. Windisch's promotional pamphlet on Kempelen was widely translated and circulated, and Joseph Friedrich Freiherr von Racknitz (1744–1818) came to the same conclusion as Decremps: that a person "of small stature" [von kleiner Statur] operated the machine with the help of magnets; Joseph Friedrich Racknitz, *Ueber den Schachspieler des Herrn von Kempelen und dessen Nachbildung* (Leipzig and Dresden: J.G.I. Breitkopf, 1789), 43. See also Robert Willis, *An Attempt to Analyze the Automaton Chess Player of Mr. de Kempelen* (Edinburgh: A. Constable, 1821).
54. Heinrich Christian Beck, *Die Schach-Maschine. Ein Lustspiel in 4 Aufzügen* (Vienna: Johann Baptist Wallishaußer, 1798).
55. Gerald M. Levitt, *The Turk, Chess Automaton* (Jefferson, NC: McFarland and Company, 2000), 44.
56. Standage, *The Turk*, 44 and 206.
57. George Allen, "The History of the Automaton Chess-Player in America," in *The Book of the First American Chess Congress* (New York: Rudd and Carleton, 1859), 420–484, 429 and Levitt, *The Turk*, 68.
58. Levitt, *The Turk*, 139 and Standage, *The Turk*, 210.
59. Metzner, *Crescendo of the Virtuoso*, 185.
60. Standage, *The Turk*, 146.
61. Standage, *The Turk*, 146 and 150.
62. Standage, *The Turk*, 113–116.
63. James Berkley, "Post-Human Mimesis and the Debunked Machine: Reading Environmental Appropriation in Poe's 'Maelzel's Chess-Player' and 'The Man That Was Used Up,'" *Comparative Literature Studies*, 41, no. 3 (2004): 356–376, 367.
64. Edgar Allan Poe, "Maelzel's Chess-Player," in *The Complete Works*, ed. James A. Harrison, vol. 14 (New York: AMS, 1965), 6–37, 35.
65. Standage, *The Turk*, 185.
66. Standage, *The Turk*, 185 and 191.
67. Additional literary depictions include the play by Jules Adenis and Octave Gastineau, *La Czarine: Drame en cinq actes et huit tableaux* (Paris: Michel Lévy, 1868) and the novel by Sheila E. Braine, *The Turkish Automaton: A Tale of the Time of Catherine the Great of Russia* (London: Blackie and Son, 1898). The Chess-Playing Turk appears in the silent movie, *White Tiger* (1923), directed by Tod Browning and a silent French film, *Le joueur d'échecs* (1927), directed by Raymond Bernard and based on a novella by Henry Dupuy-Mazuel.
68. Standage, *The Turk*, 125.
69. Johann Christian Wiegleb, *Die natürliche Magie* (Berlin and Stettin: Friedrich Nicolai, 1789), 238.
70. Johann André and Cuinet-Dorbeil, *Gesänge aus dem Singspiele, Das Automat: in einem Aufzuge* (Hamburg: J. M. Michaelsen, 1787).
71. John Francis Fetzer, "Romantic Irony," in *European Romanticism: Literary Cross-Currents, Modes, and Models*, ed. Gerhart Hoffmeister (Detroit: Wayne State University Press, 1990), 19–36.

72. Maria M. Tatar, "E.T.A. Hoffmann's 'Der Sandmann': Reflection and Romantic Irony," *MLN* 95, no. 3 (1980): 585–608.
73. E.T.A. Hoffmann, "Automata," in *The Best Tales of Hoffmann* (hereafter *BTH*), ed. with an introduction by E. F. Bleiler, trans. Major Alexander Ewing (New York: Dover, 1967) 71–103, 95.
74. E.T.A. Hoffmann, "Die Automate," in *Sämtliche Werke*, vol. 4, *Die Serapions-Brüder*, ed. Wulf Segebrecht w/collab. Ursula Segebrecht (Frankfurt am Main: Deutscher Klassiker Verlag, 2001) [=*DKV* 4], 396–427, 418.
75. Representative works include: Carl Philipp Emanuel Bach, *Stücke für Spieluhren und Drehorgeln* (Wotq. 193); Wolfgang Amadeus Mozart, *Fantasie d-Moll* (orig. f-Moll): *Orgelstück für eine Uhr*, KV 608; and Joseph Haydn, *Stücke für das Laufwerk* (Flötenuhrstücke). In addition to the piece for Maelzel, Beethoven wrote pieces for the mechanical flute ("Adagio assai für Flötenuhr, F-Dur": "Scherzo für eine Flötenuhr, D-Dur"; "Allegro für eine Flötenuhr in G-Dur"; "Grenadiermarsch für eine Flötenhuhr").
76. Robert Löhr, *The Chess Machine*, trans. Anthea Bell (New York: Penguin, 2007), 342.
77. *BTH*, 94 and *DKV* 4, 417.
78. "Der Professor—als Typ—ist bei Hoffmann stets ein Vertreter der mechanistisch-automatisch aufklärerischen Weltanschauung, die letztlich zerstörerisch wirkt"; Dieter Müller, "Zeit der Automate. Zum Automatenproblem bei Hoffmann," *Mitteillungen der E.T.A. Hoffmann-Gesellschaft*, vol. 12 (1966): 1–10, 6.
79. E.T.A. Hoffmann, "Der Sandmann," E.T.A. Hoffmann, *Sämtliche Werke*, vol. 3, *Nachtstücke, Klein Zaches, Prinzessin Brambilla: Werke 1816–1820*, ed. Hartmut Steinecke w/collab. Gerhard Allroggen (Frankfurt am Main: Deutscher Klassiker Verlag [=*DKV* 3]1985 [2009]), 3, 11–49, 24.
80. Charles E. McClelland, *State, Society, and University in Germany 1700–1914* (Cambridge: Cambridge University Press, 1980), 81.
81. McClelland, *State, Society, and University in Germany*, 170.
82. McClelland, *State, Society, and University in Germany*, 90–91.
83. E.T.A. Hoffmann, "The Sandman" in *BTH*, 183–214, 212; *DKV* 3, 46.
84. "der pedantische, exakte Kammergerichtsrat. . . . Das Satananische aber setzt er mit dem Automatischen gleich"; Walter Benjamin, "E.T.A. Hoffmann und Oskar Panizza," in *Gesammelte Schriften*, vol. 2, no. 1 (Frankfurt am Main: Suhrkamp, 1991), 641–648, here 643.
85. Walter Benjamin, "On the Concept of History," *Walter Benjamin: Selected Writings*, trans. Harry Zohn, ed. Howard Eiland et al., vol. 4 (Cambridge: Harvard University Press, 2006), 389–400.
86. Walter Benjamin, "Über den Begriff der Geschichte," in *Gesammelte Schriften*, vol. 1, no. 2 (Frankfurt am Main: Suhrkamp, 1991), 691–704, 693.
87. Walter Benjamin, "Theses on the Philosophy of History," in *Illuminations*, trans. Harry Zohn (New York: Harcourt, 1955), 255–266, 255.
88. Benjamin refers to a hunchbacked dwarf in "Berlin Childhood around 1900" and in his essay on "Franz Kafka"; see Joshua Robert Gold, "The Dwarf in the Machine: A Theological Figure and its Sources," *MLN* 121 no. 5 (Dec. 2006): 1220–1236, 1227.
89. Edgar Allan Poe, "Maelzel's Chess-Player," in *The Complete Works*, ed. James A. Harrison, vol. 14 (New York: AMS, 1965), 6–37, 19.
90. Gold, "The Dwarf in the Machine," 1227.
91. Gold, "The Dwarf in the Machine," 1228.
92. Michael Löwy, *Fire Alarm: Reading Walter Benjamin's 'On the Concept of History'*, trans. Chris Turner (London and New York: Verso, 2005), 27.
93. Löwy, *Fire Alarm*, 25.

94. Lilly Irani, "Difference and Dependence among Digital Workers: The Case of Amazon Mechanical Turk," *The South Atlantic Quarterly*, 114 no. 1 (Jan. 2015): 225–234, 227.
95. Irani, "Difference and Dependence among Digital Workers," 231.
96. Karl Marx and Friedrich Engels, *Capital,* trans. Samuel Moore and Edward Aveling (Moscow: Progress Publishers, 2015), 265; *Das Kapital*, in *Werke*, ed. Institut für Marxismus-Leninismus beim ZK der SED, vol. 23, no. 1 (East Berlin: Dietz, 1962), 401–402.
97. Andrew Ure calls Bowden's spinning loom "a powerful automaton" and another machine a "perfect automaton": *The Cotton Manufacture of Great Britain: Systematically Investigated, and Illustrated by 150 Original Figures, Engraved on Wood and Steel; with an Introductory View of Its Comparative State in Foreign Countries, Drawn Chiefly from Personal Survey* (London: H. G. Bohn, 1836), 3 and 86.
98. Paul Grimstad, "Antebellum AI: 'Maelzel's Chess-Player' and Poe's Reverse Constraints," *Poetics Today*, 31, no. 1 (Spring 2010): 107–125, 111.
99. Klaus Benesch, *Romantic Cyborgs: Authorship and Technology in the American Renaissance* (Amherst: University of Massachusetts Press, 2002), 109.
100. Poe, "Maelzel's Chess-Player," 28.
101. Voskuhl, *Androids in the Enlightenment*, 174.
102. Liu, *Copying Machines*, x.

5

RATTLED WOMEN, SHAKEN TOYS

Wollstonecraft, Baudelaire, and the Musical Lady

Erin M. Goss

If Romantic representations of automata provide ways to reflect on mechanized capital and an increasingly industrialized labor force, the earlier representations considered in this essay invite thought about a different form of labor and work as they offer figures for the operation of sexual difference through their embodied insistences on naturalized binary gender. Representations of automata that offer figures for a mechanized workforce suggest a version of humanity threatened by the encroachment of mechanized industrial systems on idealized understandings of human autonomy; they thus serve to idealize human creativity and ingenuity while holding up the automaton as a figure of horror. The earlier exhibitions and figures I will discuss here are engaged in a different sort of work. As an exploration of the kinds of figures for gendered persons that automata become, this essay focuses on the ways that late eighteenth-century figures of automata often encode not generalized humanity but specifically gendered notions of what it means to be fully human and who gets to do what to whom based on those notions. Ultimately, in the texts I consider, the automaton is not a benign figure but one that invites imagined violence and justifies acts of aggressive self-assertion at its expense. To begin, I turn to an automated woman who plays the keyboard.

Beginning on Wednesday, December 13, 1775 and during the afternoon and evening of all days but Sunday and Thursday well into 1776, London spectators could attend the Great Room at 6 King Street, Covent Garden, to witness three remarkable specimens of automated life. These three animated figures produced by the father-son team of Pierre and Henri-Louis Jaquet-Droz simulated human creative activity: one wrote; one drew; one played a keyboard instrument.[1] They also embodied and defined as natural very specific notions of gender, and especially of eighteenth-century bourgeois European femininity. That such an idea of gender is best represented by an automaton or toy is, as this essay will discuss, rather fitting, given the frequent association of women with figures of repetition and automation. What is perhaps not immediately apparent in considering the vision of gender that *La musicienne*, for that was the name given to the Jaquet-Droz harpsichordist, offered to its spectating public is the degree to which such a vision licenses acts of violence against the living women whose personhood is evacuated through association with the automaton. The automaton serves as a figure for the late eighteenth-century denigration of women, and its employment to define women's character offers an excuse for their abuse. The gendered gadget on display in London's Covent Garden thus provides a valuable starting point to consider the unfortunate association of women's character with automation rather than autonomy and with mimicry rather than will. Such an association neither originates nor concludes in the late eighteenth-century moment of this particular automaton's display. As this essay will discuss, the association of women with automata lies at the heart of Mary Wollstonecraft's late-century effort to argue for their rights as autonomous subjects. It also justifies and excuses the violence that Wollstonecraft sought to end. The allegedly lifelike manner of the Jaquet-Droz automaton belies expectations of women's behavior, as will be seen in a discussion of the appearance of automata in literature aimed at governing women's conduct in the decades following the automaton's exhibition. Ultimately, this essay will focus on what the figure of the automaton makes possible—not for women but for those who seek to control them. Lingering in the figure of the automaton is a promise of violence, aimed either to disrupt an automation that renders women disturbingly inhuman or to punish them for exhibiting an autonomy that exceeds the smooth automation of mannered behavior. Neither satisfactorily automated nor sufficiently autonomous, woman becomes the subject of violence either because she is to be punished for her insufficient humanity or because she is to be put in her place. Either way, she is, as Wollstonecraft will assert, "man's rattle," and the sound of that rattle is only amusing to the one who shakes her.

The Jaquet-Droz exhibition, called the *Spectacle Mécanique*, was neither the first nor the last such exhibition in London, and its presence in Covent Garden may have signaled its potential for unseemliness, or at least edge. Indeed, the elder Jaquet-Droz may have been imprisoned by the Spanish Inquisition for his alleged trafficking in the supernatural when he had exhibited an earlier and smaller version of an automaton at the Spanish court.[2] The exhibit had come from Switzerland, via Geneva and Paris, and would return to the Continent in 1776.[3] It was also mimicked, with replicas of the Jaquet-Droz automata appearing in London "well into the nineteenth century," according to Richard Altick.[4] While it may not be possible to determine how many people saw either this exhibition or the even more famous forays into automatic life of Vaucanson's digesting duck or von Kempelen's fraudulent chess-player, the fact of their exhibition nevertheless provides a way to think about the assumptions that govern the display of human behavior.[5] Specifically, what concerns me here as a way to begin the current essay's exploration of figures of automata as they provide license for forms of gendered violence, is what the Jaquet-Droz automata demonstrate about gender and about gender's exhibition, and the degree to which automation becomes a way to think about femininity in particular. The *Spectacle Mécanique* stages sexual difference as a category distinction somewhere between nature and culture and between authenticity and artifice. In its presentation of gendered humanoid automata, the exhibit purports to present in "art" what exists in "nature"; and yet, what it presents as natural sexual difference, of course, is produced through the artifice of cultural insistence. While it will come as no surprise to a contemporary reader to think of gender as the result of culture rather than nature, the exhibition of these particular automata provides an opportunity to witness the use of the humanoid automaton—the artificial human—as it enacts the naturalization of both gender and binary sexual difference. The automaton, that is, becomes a tool of naturalization, and its alleged imitation of human behavior becomes a means of enforcing it.

Like most exhibitions of automata and mechanical wonders in the late eighteenth century, the "*Spectacle Mécanique; Or, Mechanical Exhibition*" advertised itself not as magic or as an example of supernatural intervention but as a triumph of human achievement. The exhibition's announcement, running first in the *Public Advertiser* on the show's opening day decried, "Nature in this Exhibition is rivaled by Art," before going on to describe each of the three humanoid figures offered within it: a writer, a draughtsman, and a keyboard player.[6] That the human figures are themselves engaged in the production of art is, of course, no accident, as each engages in a kind of celebration of what human ingenuity can achieve: writing; drawing; musical performance; and governing all three, the successful duplication of human bodily movement and even, in the case of the keyboard player, the simulation of human feeling. Though the three figures are presented in terms of their productive activity and thus aligned with artistic output, the product of each has been generated elsewhere. The boys produce linguistic or pictorial written artifacts that they copy while the woman produces an art of sound composed by another; there is a book of music in front of her as if she plays from a score, and her songs were reportedly written by Henri-Louis Jaquet-Droz himself[7] (fig. 5.1). All three human figures thus offer versions of the automaton as a mimic who copies the work of another rather than generating anything through creative acts. As evidence of the way that "Nature . . . is rivaled by Art," then, they call for

Figure 5.1 Henri-Louis Jaquet-Droz, *La musicienne* with the two boys (Musée d'art et d'histoire de Neuchâtel)

the praise of human ingenuity: rather than the performers, it is the artist-machinist who has built them who is to be celebrated, and they, like their artistic output, offer mere copies of creative impulses taken from elsewhere. In their presentation as part of this larger reflection on Art's ability to rival Nature, the three Jaquet-Droz automata also offer up a version of gendered behavior that naturalizes gendered differences under the auspices of a celebration of art. In asking its spectator to revel in the ability of human creativity and ingenuity to reproduce and even rival "nature," the exhibition posits a natural femininity as a thing that is merely copied rather than produced by art and artifice.

In its inclusion along with the human figures of a "Pastoral Scene"—an automaton theatre called *La grotte* [The Grotto]—the exhibition drives home the relationship of art and nature that lies at the core of its spectatorial values. In this "Pastoral Scene" appear "a great Number of Figures; the Trees blossom and bear Fruit, the Sheep bleat, the Dog barks, and the Birds sing; each so distinctly imitating Nature that they exceed every Account that can be given of them, not only for the Variety but for the Exactness of their different Operations." Just as the viewer is invited to marvel at the degree to which the exhibition's mechanical trees replicate the actions of organic ones, so too should they be prepared to admire the degree to which the human automata on display successfully automate and mechanize human activity. The announcement's opening paragraph concludes with the assertion, "Their Mechanism surpasses every Thing that has ever appeared, insomuch that it may be strictly said they will speak for themselves." Coming after the description of the items contained within the pastoral scene, the possessive pronoun here may refer only to the trees, sheep, etc. therein; coming at the end of the entire paragraph describing the exhibition, it may refer to all items within it. The humanoid figures of the exhibition are thus framed as part of a larger scene that announces art's ability to replicate nature. However, the framing of the humanoid automata by the pastoral scene reminds a viewer that when art replicates nature through, for example, the mechanical representation of natural phenomena like the blossoming of trees and the singing of birds, it may also participate in a longstanding pastoral tradition of improving upon the nature that it describes. If art rivals nature in the pastoral, art ultimately wins, as pastoral convention replaces painful or discomfiting reality with ostensibly more desirable illusion. Christopher Marlowe would likely have little chance, after all, of convincing his reticent Love to accompany him to a working sheep farm.

The illusory idealism of the pastoral and the degree to which art is held up as superior to nature form a crucial part of this Spectacle's representation of human activity and identity, especially as it connects to and depends upon gender. The announcement begins with a matter-of-fact description of what will be on display: "Nature in this Exhibition is rivaled by Art; one Figure writes whatever is dictated to it, and draws, and finishes in a masterly Manner several curious Designs; another plays divers Airs on the Harpsichord. There is also a Pastoral Scene. . . ." Notably, none of the three humanoid automata on display is described with reference to sex, and the indefinite pronouns that denote them withhold any gendered differentiation; this, of course, makes its own kind of sense, seeing as how they are all machines and thus lack, one assumes, any necessary form of sexual differentiation. However, the actions that they perform bear such deeply gendered associations that it should be no surprise when the exhibition reveals that the first two figures described (the writer and the draughtsman) are depicted as male, while the third (the player of diverse airs on the harpsichord) is presented as female. Embodied in the three figures is an insistent claim about how gender works—not grounded in any function of lived bodies but accepted as a given well before those bodies can show up. As Adelheid Voskuhl writes in her book-length consideration of the Jaquet-Droz harpsichord-playing automaton, these automata are not "epistemically relevant simulations of human bodies, but rather mechanical reproductions of cultural and political bodily practices and ambitions."[8] These automata do not represent bodies, in other words, but ideas about bodies, and a canny reader would have known even before seeing the figures that the keyboardist was bound to be female, while the other two, unmarked by any particularly gendered activity, were likely male. While the announcement's verbal description of the automata can pretend to reserve judgment on gender, the exhibition of the automata themselves must, apparently, present gender if it is to depict "Nature," and in that presentation it can offer only the most idealized—or, in other words, the most caricatured version of gender and gendered behavior. It does so, however, as an insistence upon a "Nature" that is to be only imitated rather than improved upon by Art, even as it aligns with a pastoral scene that in its allusion to a long-standing tradition of art's triumph over nature suggests otherwise.

Even as the exhibition drew attention to the mechanical expertise of the clockmakers who had produced it, part of the experience of the spectacle was the opportunity to engage with the products generated by the automata, and in the difference in the kinds of products coming from them emerges a crucial gendered difference. The keyboard-playing automaton was remarkable for the articulation of its bodily movement, with fingers that actually depressed the individual keys of the instrument and with an internal mechanism that caused its chest to rise and fall as if moved by at least breath and perhaps even powerful emotion.[9] Described technically as "an organ mechanism with two registers and bellows," it was described by spectators in far less technical ways: "She is apparently agitated with an anxiety and diffidence not always felt in real life . . . her eyes then seem intent on the notes, her bosom heaves."[10] Intriguingly, spectators' response to her, as well as to the two boy automata, seems to have been largely governed, or at least influenced, by a pamphlet distributed by Henri-Louis Jaquet-Droz at the time of the exhibition. As Voskuhl describes, all known written responses to the exhibit repeat almost verbatim the language of the Jaquet-Droz family's own descriptions in what might be considered a Catalogue of the exhibit.[11] There, the emphasis is on the skill of workmanship that has produced the various effects on display (the goal was, after all, to sell timepieces), but the emphasized effects varied by the gender of the automaton under discussion. While the keyboard-playing automaton did in fact produce music, the pamphlet drew attention to the way that the automaton could be seen to replicate the bodily movements of a woman in the act of feeling, making the music a secondary marvel. While the boy automata produced material objects as artifacts of their mechanized endeavor, the woman produced herself, a sign of gender that could be taken as the proper object of her spectacular presentation.

The output of the two boy automata was, logically enough, the material product of their acts of writing, but that of their female counterpart was not only the music it played but also its automated performance. What the female automaton produced was affect and, indeed, the ironic insistence it offered of a natural form of gender. The writer automaton copies a sentence, being so meticulous as to return once it has reached its period to dot the *i*'s and cross the *t*'s. The draughtsman, similarly, fills several sheets of paper as it ostensibly copies a drawing that has been placed in front of it. In both cases, the evidence of their mechanized work appears in papers that can be distributed to an audience once they have been filled by the mechanical actions of the automaton. The harpsichordist, by contrast, produces no material artifact, and description of it as "The Musical Lady" focused on its breast and on the signs of its emotion, usually using a feminine pronoun. There is evidence that spectators were able to take home with them examples of the writing and drawing completed by the two boy automata, thus providing tangible evidence of their pseudo-artistic endeavors.[12] What spectators could take home from *La musicienne*, by contrast, was their confirmed sense of how women function.

If the figures offer explicit comment on the relationship of art and nature, as their presentation along with the "Pastoral Scene" insists they do, they also provide a more implicit though equally direct comment on the ostensible fact of gender. After all, it is worth noting that while all three figures are engaged in acts of copying and mimicry that announce their lack of access to creative impulses of their own, the two male figures are offered up as representations of children while the female figure is an adult, or at least nearly one. The boys' work is that of the copyist, but so too was the work of the male pupil, who learned the art of writing through acts of emulation that afforded an eventual education first in penmanship and then in composition more properly.[13] The copying boys represent a stage in the education of men, and their copying thus stands in for a copying that will eventually (or would, if these were "real" boys) give way to a more creative and individualized expressivity. The boy automata, that is, represent an only temporary developmental status in which one is automated *en route* to becoming not so. *La musicienne*, on the other hand, presents a figure of an adult woman rather than a pre-pubescent boy. Though the brochure distributed at the exhibition described *La musicienne* as "a young girl between ten and twelve years old," the automaton herself clearly represents a more mature female figure in facial and anatomical features (fig. 5.2). While certainly designed to represent a "diffidence" and perhaps even a youthful innocence, the female figure's face depicts at the youngest a juvenile, and the emphasis on her heaving breast places her closer to the position of adulthood than the announcement itself seems to mark. *La musicienne*, unlike her puerile counterparts is not, or at least not fully, a child, and in contrast to theirs, hers does not depict a temporarily automated status in the progress of her education but rather that education's end result.

Figure 5.2 Henri-Louis Jaquet-Droz, Close-up of *La musicienne*'s face (Musée d'art et d'histoire de Neuchâtel)

Figure 5.3 S. W. Fores, *Lady Squabb Shewing Off, or a Punster's Joke*, 1811 (Lewis Walpole Library, Yale University)

It is no accident that *La musicienne* is playing a keyboard instrument, as women's association with such instruments was ubiquitous enough to warrant numerous parodies, as well as more traditional representations across media.[14] However, women's association with keyboard-playing often had little to do with music. Musical instruction was central to women's education insofar as that education primarily prepared them for courtship and marriage. As Regula Trillini describes, "Between the sixteenth century and the Great War . . . music featured in every plan for female education," and music "was consistently prescribed for women, with the obvious, though not always explicit, goal of attracting suitors that had been brought up to despise the art but were conditioned to consider it an erotically stimulating spectacle."[15] Men in the eighteenth century, by contrast, rarely received an education in music, and it was considered at best a distraction. John Locke, for example, dismisses its appropriateness as a subject for boys on account of the time it takes away from more worthwhile pursuits, claiming that among its detriments, it "wastes so much of a young Man's time, to gain but a moderate Skill in it," not to mention that it also "engages often in such odd Company." Furthermore, Locke reports that he has "amongst Men of Parts and

Business, so seldom heard any one commended, or esteemed, for having an Excellency in *Musick*" that he, in short, "give[s] it the last place" on the list of things that might be studied. (Dancing, incidentally, "cannot be learn'd too early.")[16] As Richard Leppert writes in his study of representations of musicians in eighteenth-century England, "most writers of conduct literature agreed that music, broadly understood, was essentially ungentlemanly; music was improper for a man because it was unmanly," and "the visual semiotic of men performing music is one where power has no play."[17] Not only does the physicality of the act of playing result in nothing lasting or measurable (especially in an era before widespread recording technology), but it also renders the performer the object of a spectatorial gaze that rarely puts men in its sights. The musician leaves behind nothing tangible, producing only airy nothings and the ephemerality of a moment of being observed in performance. For women, the performance of music is an opportunity to demonstrate the results of an education premised upon practiced repetition; and performance upon the keys allows for the display of proper posture and disciplined demeanor. The keyboard player herself, as mocked in caricatured figures like "Lady Squabb," is the object of visual attention more than she is a producer of music, and the point of playing music is to make oneself available as a visual object rather than to aspire to any great skill or even, dare one say, accomplishment (fig. 5.3).[18]

Perhaps because it provides license for exhibition, music was an essential component of women's education in preparation for the marriage market. Its purpose and function, however, was far from aesthetic, at least insofar as it was discussed from the perspective of observers or any potential audience. Girls were trained to perform, but that performance was to occur under very tightly controlled circumstances. Indeed, Leppert tells the story of Ann Ford, the subject of a well-known portrait by Thomas Gainsborough, whose father had her arrested so that she would stop providing public concerts.[19] The training in music received by girls in the eighteenth century aimed at the specific purpose of courtship. As a character in George Colman's *The Musical Lady* (1762) pronounces: "I dare say this passion for music is but one of the irregular appetites of virginity: You hardly ever knew a lady so devoted to her harpsichord, but she suffered it to go out of tune after matrimony."[20] And it should come as no surprise that one of the most insufferably artificial and showy of Jane Austen's characters is not only marked as "a superior performer" but also prides herself on being "doatingly fond of music—passionately fond" even as she imagines a future in which she, like numerous of her acquaintance whom she mentions, will have "entirely given up music."[21] For the imagined female performer, the production of music is a secondary goal, a merely necessary side effect of musical performance, as the creators' own emphasis on *La musicienne*'s heaving breast may attest. The observer of the mechanical spectacle of a female figure playing a harpsichord does not marvel at the articulation of the automaton's fingers nor at the fact that the machine produced actual music. Rather, what warrants remark is the sense that the figure is "apparently agitated with an anxiety and diffidence not always felt in real life." That is, not only does the automaton replicate for an audience the apparent emotional output of a living woman; it may in fact surpass her apparent authenticity.

If the Jaquet-Droz automaton embodies social codes regarding the performance and exhibition of gender, those codes themselves often rely on the more general figure of the automaton for their articulation. *La musicienne* thus serves as a figure (rather literally) for a femininity under construction throughout the eighteenth century in conduct manuals and popular novels alike. There, the automaton is a figure of both horror and aspiration, though rarely in an acknowledged way. Julie Park writes about the ways that the automaton becomes "a model of mimesis and regularity" within "eighteenth-century literature and social life."[22] Park offers up the figure of the automaton as what she calls an "artefactual corollary" for women's social identity in its "mixed figure for women's consistency and restraint, or her vanity and vacuity." Indeed, if the automaton appears as a figure for what women might become through their education, it also appears in both conduct manuals and novels as a bad model, an object lesson in what the author of the 1774 "Letter from a father to his daughter at boarding-school" laments as the primary problem of women's education. The education that a young woman receives at boarding school, according to this nobly directed and concerned father, produces a training in "mechanical things, in which the head has the smallest share."[23] The version of femininity aimed at and produced by such an education promises evenness and adherence to expectations; properly trained women do not surprise. Such an education is offered up not as a training that overcomes a woman's natural inclinations but as the opportunity to more fully inhabit them. As the doting father himself remarks, "A woman ruffled with passion makes a disagreeable appearance; *indeed she is not herself*."[24] Passion, or any expression that would single one woman out from another, renders her not only unappealing in this assertion but in fact unnatural and inauthentic.

While the father writing to his daughter at boarding school in 1774 might lament the mechanical nature of his daughter's education, other primers would claim that a mechanical education is appropriate for members of a sex deemed to be themselves nearly mechanical. As Thomas Gisborne remarks in his introduction to *An Enquiry into the Duties of the Female Sex* (1797), women are fundamentally interchangeable, subject to what he deems "the general similarity of women"; while they may differ slightly in terms of outward circumstance, there is little to distinguish them from one another in character.[25] After all, as Alexander Pope remarked in 1743, "Most Women have no Characters at all."[26] For Gisborne, as for many of his fellow pamphleteers in the arena of women's education, women are to be taught to be replicable: "they are not to envy, but to admire; to copy, not to emulate" (79). Ultimately, the goal of such an education, according to Gisborne, should be "a well-regulated life" (236) governed by the kind of rules and habits that make for slim distinction between life and automation. In fact, it seems that for Gisborne and others a mechanical education of automated repetition is appropriate and even natural to the female half of the species; as he remarks, the "propensity to imitation . . . natural to the human mind . . . shews itself with especial strength in the female sex" (123). Women's education, so goes the logic, is a training in automation because women are themselves barely distinguishable from mechanized automata.

It is just this sort of mechanical education that leads Mary Wollstonecraft to call in her *Vindication of the Rights of Woman* for "a revolution in female manners."[27] For Wollstonecraft, it is not the specificity of which "manners" women are taught that causes them to live without "dignity" but rather that "manners" define all that they do. Women's education provides training in a "false system" of habits and assumptions that denies women an education tending toward understanding in exchange for indoctrination that promises access to power through the successful manipulation of sexual expectations, as women are taught to use the "wit and cunning" available to them as sexualized objects to gain what small rudiments of control and apparent autonomy they might be able to grasp (167, 168). Not unlike the Jaquet-Droz automaton, Wollstonecraft's female subject draws upon her heaving bosom and displays of ostensibly human feeling to gain a fleeting and illusory form of attractive power. Whether or not she can play the harpsichord, speak a sentence, or actually feel seems to be rather beside the point.

Indeed, Wollstonecraft underscores throughout the *Vindication* the degree to which women's education has created them as mechanisms for the gratification of others. Her contemporary fellow woman, she exhorts, "was created to be the toy of man, his rattle" (144). And, she goes on, as rattle, "it must jingle in his ears whenever, dismissing reason, he chooses to be amused" (144). Wollstonecraft's lament over woman's degraded status succinctly pinpoints the very problem in women's education. As a toy, woman becomes a thing, and as a thing, she acquires the neuter pronoun "it."[28] In the slippage of pronoun emerges the heart of Wollstonecraft's argument pertaining to women's education in manners. For Wollstonecraft, to become a toy is to lose human identity in the acquisition of automation. There is, however, something ironically hopeful in such an insistence that should not be missed: those who have been made to be toys are not, in fact, women, for as soon as they are created as toys they cease to be identifiable as anything other than objects. Wollstonecraft historicizes what her contemporary writers and other producers of automatic women have naturalized. These women whom she criticizes (rather harshly, as has been pointed out[29]) act as toys, but they do so as a result of their creation not by nature or any version of divine authority but by a system of education that values their automation and interchangeability over their humanity. Such a system is subject to change. At stake for Wollstonecraft is an imagined and as yet unknown ontological status: women could perhaps be women if they were not in the state of having been and consistently being again created as toys. However, because women are persistently created and re-created as toys, as Kirstin Wilcox states succinctly in her reading of the *Vindication*, "women themselves remain unknown."[30] Identified only by a feminine pronoun that vanishes into the ungendered "it" of the toy, women are subsumed in the figure of the automated creature as which and through which they are imagined.

If Wollstonecraft's pronoun shift makes imaginable a future in which women can come to be women rather than mechanized playthings, her passive verb both deflects blame and underscores the fuzziness of the line between the natural and the artificial in women's identity. It is unclear from Wollstonecraft's sentence who it is that is effecting this apparently constant creating and re-creating of women as toys. If women have been created as toys, then by whom? And, indeed, when? Wollstonecraft's text takes care not to lay the blame at the feet of a creator and thus deem women's artifice and automation the result of divine fiat or the stuff of nature. She even spends much of the beginning of her text criticizing not only Milton but also Moses for suggesting as much, as she identifies the ideological usefulness of an assertion of God as toy maker and women as toys to those who want their toys to jingle for them on the regular. Ultimately, the identity of the creator of woman-as-toy remains as unknown as the identity of the toy itself; lost in the grammatical indeterminacy of a passive verb construction and a pronoun that violently severs its relationship to its own antecedent, both toy and creator are rendered strangely inhuman—or at least un-personed. This deflection of blame may be a rhetorical strategy; after all, a large part of Wollstonecraft's audience comprised the men who might well be currently engaged in making their toys jingle for them.[31] For that portion of her audience, Wollstonecraft may insert the subtle reminder that when "it" (the rattle or toy) jingles, it does so because some undisclosed and unnamed "he" has "dismissed reason" to ask it to do so. Given that reason is, throughout the *Vindication* and the intellectual world into which it asserts itself, the very thing that defines humanity (and, indeed, that often purports to excuse the hierarchical division of the sexes), this is no accidental aside.

There is no indication that Wollstonecraft saw the exhibit of the Jaquet-Droz automata. Indeed, the five shilling price would almost certainly have rendered the

Spectacle Mécanique outside the experience of a seventeen-year-old Mary Wollstonecraft. The argument I make here, however, does not seek to ascribe causation or even influence as a relation between the two figures—the harpsichord-player and the toy rattle. Rather, I seek to show that the version of femininity that Wollstonecraft identifies as impermanent and historical, and thus changeable, the Jaquet-Droz exhibition embodies as natural and thus unquestioned. The femininity that grounds the presentation of the objects that Voskuhl calls "luxury commodities" is inseparable from that which Wollstonecraft denigrates and protests in the "toy" her fellow woman is taught to be. While the heaving bosom of the former has the apparent goal of selling clocks (the Jaquet-Droz name is still associated with extremely high-end timepieces), one might well ask what purpose is served by the toys that breathing women are trained to be.

Wollstonecraft identifies late eighteenth-century femininity as the result of an education that promises to produce women as properly mannered beings. Such an education is grounded in the tenets of Jean-Jacques Rousseau, to whom Wollstonecraft devotes a good number of pages in justifiably angry rebuke; James Fordyce, that object of Jane Austen's none-too-subtle mirth in *Pride and Prejudice*; and John Gregory, whose *Legacy of a Father to His Daughters* (1774) Wollstonecraft describes as a call for an initiation into a "system of dissimulation" despite its admirable paternal affection. Quoting extensively from the authors she derides, Wollstonecraft demonstrates that the education in sentiment and the pleasure of others promoted by these most respected authors results in women who excel in the "display of cold artificial feelings" (218) rather than any kind of proper sensibility, let alone knowledge or understanding. Such doctrines render women an apparent "defect in nature" that must be righted by an education that trains proper behavior because reasoned action is deemed impossible; they imagine women as requiring the automation of habit in order to function in a social world. They in effect aim to create women as automata.

While certain dominant strains of women's education aim to produce women as automata, concurrent strains of thought—often within the same publications calling for women's training through habit and manners—decry the degree to which women might be too automatic. Thus, even as the figure of the automaton embodies the logical consequences of women's education, it also becomes a sort of scarecrow in novels of the late century, as well as in certain conduct manuals. The automaton is a trap into which women might fall by having been successfully taught to be the thing that their education seemed destined to produce. After all, as is often suggested, the mechanical education she receives at boarding school might train a woman right out of what makes her alluring in the first place. In Mary Barker's *A Welsh Story* (1798), for example, a male character deigns to teach the novel's heroine how to evade the pitfalls of a women's culture that would lead her toward apparent automation. Sir Edwin explains, "[T]he whimsicality of manner, which many women possess, together with a little eccentricity, is frequently the secret mark of genuine goodness of disposition, sprightliness of imagination, and unconquerable simplicity. If the manners of such women deserve to be called affected, I most sincerely wish that all my acquaintance might be so; and that all women, instead of making mere automaton figures of themselves, and polishing till they have no mark of rationality about them, rather entertain us with the follies of nature, than of art" (182–183).[32] In this moment, a male character aims to show a female character the pitfalls of her education, which in training her to be appropriately feminine has taken from her the traits that he deems to be the more valued characteristics of femininity. Like Robert Herrick, whose "Delight in Disorder" echoes throughout the centuries into what these days might be called the Manic Pixie Dream Girl,[33] Sir Edwin prefers to be "entertained" by the apparent intrusion of "nature" into the artifice that is a woman's manners. Striking about this passage in which Barker allows Sir Edwin to remind her reader that no one seems to want a robot (perhaps a more debatable claim these days) is the suggestion that it is Charlotte who has made of herself an automaton. It is thus Charlotte who should will herself to present instead the apparent accidents of nature that will manifest as unplanned whimsy and eccentricity. As an automaton, she is of no interest. As an automaton that fails, or appears to fail, to remain automatic, she becomes appealing again and she again possesses the power to "entertain."

Other women novelists identify the automaton as a means to denigrate women.[34] In Mary Charlton's *Rosella, or Modern Occurrences* (1799), the term becomes a way to dismiss, with some violence, a woman who has not fulfilled expectations, though those expectations had little to do with the woman herself, as a former suitor "declared, upon the refusal of her barbarous brother to make up his affairs, that a pale, prim, simpering, philosophical, deistical, botanical, astronomical, sophistical, soft-mannered, flint-hearted automaton was his supreme aversion."[35] Perhaps the adjectives ascribed to this young woman are appropriate, if they can even be said to cohere, but it is the noun that is the most damning. To dismiss this woman there is little more that needs to be said than that she has fallen into the trap that her education has set for her and become an automaton herself. Similarly Elizabeth Hervey's 1788 *Melissa and Marcia; or the Sisters* has one of its characters warn another: "Depend upon it, he considers you as a pretty idol, just formed for him to parade about; an amusement for a leisure hour, but with no more intelligence than an automaton."[36] Such an insistence is offered as evidence that the gentleman in question cannot possibly love the addressee.

Even Rousseau, in stipulating the kind of education that would suit the partner for his ideal pupil, Emile, insists that Sophie's intellectual deprivation should not be total. Rousseau describes at some length the degree to which women's education should be, as Wollstonecraft singles out, "always relative to the men" and that they should be taught primarily "[t]o please, to be useful to us, to make us love and esteem them," and so on.[37] However, this hierarchy of function does not, evidently, mean that Sophie should be "raised in ignorance of everything"; Rousseau asks, apparently rhetorically, "Will man turn his companion into his servant? Will he deprive himself of the greatest charm of society with her? In order to make her more subject, will he prevent her from feeling anything, from knowing anything? Will he make her into a veritable automaton?"[38] Though Rousseau answers his own ostensibly rhetorical questions emphatically in the negative, Mary Wollstonecraft, in this instance as in so many others a better reader of Rousseau than Rousseau himself, will recognize the education of Sophie for what it is: a toymaker's shop.

Young women, like Sophie, Charlotte, the woman with the disappointing brother, the daughters at boarding school, are adjured to resist the pull to automation, as if one becomes automated by choice. However, simultaneously, they are taught to be automatic and trained to display the sentiment that is also recognizable in the Jaquet-Droz automaton. Such an education can take the shape of automation precisely because woman is herself often deemed to be a copy in the first place. In the truly remarkable 1739 pamphlet "Man Superior to Woman; or, a Vindication of Man's Natural Right of Sovereign Authority over the Woman," to cite one example among many horrible others, an anonymous (and well might he be) author proclaims that women's "impotent Eagerness to be like *Man* serves only to shew, that they are but mere mechanic Rote-repeaters of *his* Words and unsuccessful Mimics of *his* sense."[39] The argument here, of course, is that women are trained through acts of automation to become automated creatures because, evidently, they are fit only to be automated and mechanical repetitions and thus serve best as toys for those whose authenticity and originality is without question. The evidence for "Woman's" status as merely unoriginal repetition is her desire to be "like Man," her desire, that is, to be recognized as an autonomous being. Such a desire is understandable only as an aim to repeat. The absurd circularity of such reasoning reveals itself quickly and arrives at its own culminating profundity in the advice, in novels like Mary Barker's, that women should do what they can to cultivate behavior that keeps them from seeming too artificial and automatic. As both the thing that women are trained to be and the thing that they are most blamed for being, the automaton comes to embody the double bind of late eighteenth-century bourgeois white femininity, and "woman" becomes nearly synonymous with the mannered toy that Wollstonecraft laments. Wollstonecraft's call for a "revolution in female manners," is, after all, not a call to change or substitute manners but rather to reject the very tendency to hold up manners—repetitive, automatic actions—as a "painted substitute for morals." For in their mastery of manners women fall prey to the double bind that the automaton represents. Trained up to have identifiable and replicable emotions, women's affect can be reproduced in the heaving breast of a machine body (like *La musicienne*). Women, however, are to be blamed for being too automatic, while the automaton is praised for being so successfully womanly.

In what is usually translated as his "philosophy" of toys, Charles Baudelaire defines the difference between the sexes through a differential relation to what we can here identify as automation.[40] As for Wollstonecraft, women in Baudelaire's schema are governed by the mindlessness of habit while men are encouraged to assert their wills. Though Baudelaire's exploration of this topic comes more than half a decade after Wollstonecraft's, his consideration of explicitly gendered relations to toys can shed a good bit of light on Wollstonecraft's suggestive use of the figure of the toy, and his reflections on toys and gender can be seen as the logical extension of the femininity produced and reproduced through the eighteenth-century figure of the female automaton.

Baudelaire's basic claim begins with what he calls "the overriding desire of most children" to "*see the soul* of their toys."[41] In efforts to fulfill this desire, children twist, turn, shake and bump their toys until they effectively destroy the pretense of life that had instigated their curiosity in the first place, as the desire to understand and possess the secrets within the toy initiates the child into a form of violence and destruction. In their "play" with these toys children develop their imaginative and intellectual capacity through a desire to know that leads to an act of revelation that is also an act of dismemberment. Not unlike William Wordsworth's fictional addressee in "Expostulation and Reply," the child "murder[s] to dissect," and for Baudelaire the desire to know through destruction brings with it a complicated set of emotions that makes the desire to know and the desire to destroy concomitant with a "mania" that could also be called love. Disappointingly enough, once the child has investigated the toy sufficiently to arrive at the potential source for the life it had evinced, the child discovers nothing. "But *where is the soul*," Baudelaire's imagined child repeats in the wake of the life he has imagined and then destroyed.[42] The introduction to melancholy and gloom that Baudelaire reports to coincide with the child's failure to locate the meaning of life in his disassembled toy initiates the child into metaphysics, philosophy, and abstract thought as such. It also establishes the parameters of lived sexual difference and defines the relation between the sexes by violence.

Unsurprisingly, when Baudelaire writes that "all children talk to their toys," he does not mean "all children," and he quite emphatically distinguishes the kind of play with toys that leads to speculative wonder from another version of play defined by mindless repetition.[43] Some children, according to Baudelaire, do not act on a destructive imagination that first animates and then dismembers their toys; they instead imitate, and their acts of mimicry strand them in a state of perpetual childishness. Perhaps unsurprisingly, this difference is decidedly gendered, and the play of "those little girls who play at grown-ups" by imitating their mothers and other women is denigrated as being hardly play at all.[44] In their play, girls prepare for a lifetime of habit. They rehearse a future as automata in which their greatest ambition is neither imaginative nor speculative but strictly mimetic. While girls are training as automata, of course, their boyish counterparts enter the intellectual world by means of violent acts that destroy the automata that they come to discover were only animated in the first place through their own acts of imagination. It is not entirely clear why the automaton must be destroyed. The boy-child's violence may come merely as a secondary consequence of his effort to understand. However, in Baudelaire's description it seems to carry a bleaker explanation. In his destruction of his toy, the boy learns his own superiority over it; he learns that he has will while the toy does not. His violence and its consequences become the evidence of his autonomy, and it doesn't seem to matter much what other cause might have lain behind it.

I am drawn in conclusion back to Wollstonecraft's identification of woman as toy, made to rattle for the pleasure of another. Indeed, it is upon this identification that Wollstonecraft's entire call for rights rests. As the entirely antagonistic reader for the *Critical Review* summarizes in the first review in which that publication condemned the book, Wollstonecraft's entire text is "an indignant invective against treating women merely as toys."[45] While this rather long and itself indignant review sputters against such a claim as unwarranted and silly, it also upholds an insistence on women's intellectual inferiority. Thus its apparent refutation of Wollstonecraft's identification of woman as toy ends up being instead a defense of such an identification. Ultimately, while claiming in its concluding gesture to undermine Wollstonecraft's argument that women are treated as toys, it instead concludes by arguing that women should be treated as toys, and it does so while also claiming to refuse to participate in the "petty warfare" of calling out "Miss Wollstonecraft's" numerous examples of "inconclusive reasoning" and logical "inconsistency."

What the *Critical Review* fails to understand—or perhaps understands all too well—is what is at stake in the identification of woman as toy. Wollstonecraft's *Vindication*, while certainly a call for an equality that might be made manifest in educational and perhaps even eventual vocational opportunity, is not merely a defense of an abstraction. It is a call against a state of affairs that licenses and excuses violence at every turn. The "toy" with which Wollstonecraft identifies her contemporary women is a rattle, an object that makes noise and produces "amusement" and is in fact only identifiable as itself when it is shaken, and the more violently the more impressively. The daughter of a persistently beaten mother and the author of *Maria; or the Wrongs of Woman*, a novel that presented the state of even the most bourgeoise and protected of women in late eighteenth-century England as that of a "misery and oppression" under constant threat of legally invisible violence, Mary Wollstonecraft's call to differentiate women from toys is a call to stop excusing and inviting violence against them.

One doesn't want, of course, to blame the Jaquet-Droz automaton for violent acts against living women. Such an accusation would be misplaced in any number of ways. What I mean to suggest, however, is that in its presentation of the ideal woman as representable through replicable affect, *La musicienne* offers a vision of living women as themselves automata, toys that bear evidence of their training in feelings and expected behavior. The difference is, of course, that *La musicienne*, having never lived, can survive her presentation as toy unscathed.

NOTES

1. The instrument played was a *clavecin*. Because this instrument was functionally very similar to the harpsichord, which has more immediate cultural resonance in an English context (and because *clavecin* and *harpsichord* are often translations for one another), I will be referring to the instrument as a *harpsichord*.
2. Sources disagree on the veracity of this anecdote, with some attesting that the elder Jaquet-Droz had merely feared imprisonment and held it off through careful explanation of the workings of his mechanical wonder. The report of the imprisonment appears anecdotally in, for example, Henry Ridgely Evans, *The Old and the New Magic* (Chicago: Open Court Publishing Company, 1909), 400–402, as well as other places. Evans offers an extended description of the two male automata included in the exhibit, even offering a sample of the artifacts produced by each. These samples—a bit of writing and a drawing—are held up as evidence of the continued functionality of each of the automata, a sign of the superior craftsmanship of Henri-Louis Jaquet-Droz, who could produce automata that would perform faithfully over centuries. Evans says almost nothing about the female figure, the keyboard player.
3. Adelheid Voskuhl claims in her extensive consideration of this exhibition that there is no evidence that it toured as extensively as others—namely Georg Schindler in his "Die legendären Automaten von Jaquet-Droz, Vater und Sohn," and Richard Altick in his *The Shows of London*—have claimed. See *Androids in the Enlightenment: Mechanics, Artisans, and Cultures of the Self* (Chicago: University of Chicago Press, 2013), 63. Voskuhl's remarkably detailed and thorough consideration of the two automata is crucial for the current essay, though she does not provide much discussion of gender or sexual difference, as is also observed by Katherine Crawford, in her discussion of the Jaquet-Droz automaton in relation to what she calls eighteenth-century "sentimental performativity," "Designed

Women: Gender and the Problem of Female Automata," *History and Technology* 30, no. 3 (2014): 261–268, 263.

4. Richard D. Altick, *The Shows of London* (Cambridge: Belknap Press, 1978), 66.
5. For an extended discussion of the eighteenth century as the "golden age of the automaton," see Minsoo Kang, *Sublime Dreams of Living Machines* (Cambridge: Harvard University Press, 2011), which offers an extended analysis of the kinds of philosophical meaning automata were understood to embody across the century. Kang, interestingly, pays little attention to gender in this analysis, though he takes up the subject from a slightly different angle in his "Building the Sex Machine: The Subversive Potential of the Female Robot," *Intertexts* 9, no. 1 (2005): 5–22.
6. This announcement first appeared in *The Public Advertiser* (originally *The London Daily Post*) of Wednesday, December 13, 1775. I am grateful to Duncan Jones at the Bodleian Library for dating the announcement's original appearance.
7. Voskuhl, 59.
8. Voskuhl, 21.
9. See Mary Hillier, *Automata & Mechanical Toys* (London: Bloomsbury, 1976), 56. Hillier singles out the Jaquet-Droz automata for their superior craftsmanship, evidenced by the fact that all three remain in working order, preserved and maintained in the Musée d'art et d'histoire in Neuchâtel (MAHN).
10. Altick, 66, 67.
11. As Voskuhl, 39, argues, because these descriptions merely transcribe language from the original text produced by the Jaquet-Droz family, "Descriptions of the automata in these print media are therefore practically identical to each other, and they do not tell us about individual people's experiences of the automata or reactions of groups of people during automaton showings."
12. Voskuhl, 65, describes, for example, the inclusion of some of the draughtsman's drawings in a letter written by one of the exhibit's early spectators.
13. Though more a medieval practice than an eighteenth-century one, the act of copying and memorization as a mode of instruction persisted well into the nineteenth century. For further consideration, see M. O. Grenby's *The Child Reader, 1700–1840* (Cambridge: Cambridge University Press, 2011), 233, in which he describes the ways that pedagogical theory began to shift away from memorization by the eighteenth century and yet concludes, "Regardless of what was being recommended . . . there is much more evidence of the continuation of rote learning into the nineteenth century than its abeyance."
14. There were in fact numerous musician automata produced in the eighteenth century, and they were not all depicted as female. There were, however, no male keyboard-playing automata.
15. Regula Hohl Trillini, *The Gaze of the Listener: English Representations of Domestic Music-Making* (Amsterdam: Rodopi, 2008), 4.
16. John Locke, *Some Thoughts Concerning Education*, sect. 196, 197 (Oxford: Clarendon, 2000), 252.
17. Richard D. Leppert, *Music and Image: Domesticity, Ideology and Socio-Cultural Formation in Eighteenth-Century England* (Cambridge: Cambridge University Press, 1988), 24–25.
18. S. W. Fores, *Lady Squabb Shewing Off, or a Punsters Joke* (6 September 1811, no. 50, Piccadilly). There are many examples of such caricatures, and The Lewis Walpole Library has several in their collection of satirical prints. See also James Gillray, *Farmer Giles & his Wife shewing off their daughter Betty to their Neighbours, on her return from School* (1809) or M. Darly, *The accomplish'd maid* (21 May 1778).
19. Leppert, 40.
20. *The Dramatick Works of George Colman*, 4th volume (London: T. Beckett, 1777), 72. Quoted in Leppert, 44.

21. Jane Austen, *Emma*, ed. Richard Cronin and Dorothy McMillan (Cambridge: Cambridge University Press, 2005), 298–299.
22. Julie Park, "Pains and Pleasure of the Automaton: Frances Burney's Mechanics of Coming Out," *Eighteenth-Century Studies* 40, no. 1 (2006): 23–49, 23.
23. Anonymous (London: Printed for G. Robinsonn, 1774). For more on the role of machines in discussions of women's education, see Brandy Lain Schillace, "Reproducing Custom: Mechanical Habits and Female Machines in Augustan Women's Education," *Feminist Formations* 25, no. 1 (2013): 111–137. For a parallel discussion of the ways that the automaton can become a figure for certain men as well, see Alex Wetmore's discussion of the automaton in relation to the masculine subject of sensibility in "Sympathy Machines: Men of Feeling and the Automaton," *Eighteenth-Century Studies* 43, no. 1 (2009): 37–54.
24. "A Letter from a father," 15, emphasis mine.
25. Thomas Gisborne, *An Enquiry into the Duties of the Female Sex* (London: T. Cadell, 1797), 3.
26. Alexander Pope, "Epistle II: To a Lady," *The Poetry and Prose of Alexander Pope*, ed. Aubrey Williams (New York: Houghton Mifflin, 1969), 167.
27. Mary Wollstonecraft, *The Vindications: The Rights of Men and The Rights of Woman,* ed. D. L. Macdonald and Kathleen Scherf (Peterborough, ON: Broadview, 1997), 158.
28. For a compelling discussion of pronouns with implications for the linkage of pronoun gender to the status of object, see Barbara Johnson, *Persons and Things* (Cambridge: Harvard University Press, 2008), 6–7. Her discussion of Barbie is also relevant to the current discussion. See 163–175.
29. As Claudia Johnson summarizes neatly, Wollstonecraft's "censoriousness of women as well as her commitment to ostensibly masculinist, enlightenment values have disappointed those who expect feminism to produce, as she does not, a positive culture of the feminine and of female solidarity," *Equivocal Beings* (Chicago: University of Chicago Press, 1995), 23.
30. Kirstin Wilcox, "Vindicating Paradoxes: Mary Wollstonecraft's 'Woman,'" *Studies in Romanticism* 48, no. 3 (2009): 447–467, 449.
31. See Anca Vlasopolos, "Mary Wollstonecraft's Mask of Reason in *A Vindication of the Rights of Woman*," *Dalhousie Review* 60, no. 3 (1980): 462–471, for an early version of the argument that Wollstonecraft wrote only for men, and Amy Elizabeth Smith, "Roles for Readers in Mary Wollstonecraft's *A Vindication of the Rights of Woman*," *SEL* 32, no. 3 (1992): 555–570 for a formal analysis seeking to identify Wollstonecraft's audience by sex.
32. Mary Barker, *A Welsh Story in Three Volumes* (London: Hookham and Carpenter, 1798), 182–183. For a reading of Barker's novel that finds it much aligned with Wollstonecraft, see Mary Chadwick, "Making Space for Wollstonecraft: Mary Barker's *A Welsh Story*," *Romantic Textualities: Literature and Print Culture, 1780–1840* 22 (Spring 2017): 21–35.
33. See Nathan Rabin's 2014 *Salon* article, "I'm sorry for coining the phrase 'Manic Pixie Dream Girl,'" accessed March 24, 2019, http://www.salon.com/2014/07/15/im_sorry_for_coining_the_phrase_manic_pixie_dream_girl/. In his original 2007 review of *Elizabethtown*, a film perhaps remembered only for its occasioning the first use of such a profoundly useful phrase, Rabin remarked of the Manic Pixie Dream Girl, appropriately enough for the current argument, that "[a]udiences either want to marry her instantly . . . or they want to commit grievous bodily harm against [her]," *A. V. Club* January 25, 2007, accessed March 24, 2019, http://www.avclub.com/article/the-bataan-death-march-of-whimsy-case-file-1-emeli-15577.
34. As Andrea Haslanger describes, the automaton is also crucial to Frances Burney's *Camilla*, in which the "mechanical behavior" that nearly all women in the novel perform at one point or another indicates the way in which "the automaton shifts from being an example

of negative femininity to being an example of femininity as such." For Haslanger, as for Burney, the automaton offers a way to name women's inauthenticity rather than acknowledging the numerous ways that women's autonomy is blocked and denied. "From Man-Machine to Woman-Machine: Automata, Fiction, and Femininity in Dibdin's *Hannah Hewit* and Burney's *Camilla*," *Modern Philology* 111, no. 4 (2014), 788–817, 814. See also Chapter 7 (on Burney) of Claudia Johnson, *Equivocal Beings* (Chicago: University of Chicago Press, 1995), 142–165.

35. Mary Charlton, *Rosella, or Modern Occurrences. A Novel in Four Volumes* (London: Minerva Press, 1799), 305.
36. Elizabeth Hervey, *Melissa and Marcia; or the Sisters: A Novel. In Two Volumes* (London: W. Lane, 1788), 125.
37. Here I use Wollstonecraft's quotation of Rousseau from Wollstonecraft, 200. A relevant passage in Bloom's translation: "woman is made specially to please man. . . . If woman is made to please and to be subjugated, she ought to make herself agreeable to man instead of arousing him. Her own violence is in her charms. It is by these that she ought to constrain him to find his strength and make use of it. The surest art for animating that strength is to make it necessary by resistance." *Emile, or On Education*, trans. Allan Bloom (New York: Basic Books, 1979), 358. Woman's use, then, is to encourage men to use their strength against them; their function, in addition to or as part of pleasing, is to be objects of violence.
38. In Bloom's translation, p. 364. In French: "S'ensuit-il qu'elle doive être élevée dans l'ignorance de toute chose, et bornée aux seules fonctions du ménage? . . . En fera-t-il un véritable automate? Non, sans doute." Jean-Jacques Rousseau, *Émile ou de l'éducation*, ed. François and Pierre Richard (Paris: Classiques Garnier, 1964), 454.
39. *Man Superior to Woman; or, A Vindication of Man's Natural Right of Sovereign Authority over the Woman* (London: T. Cooper, 1739), 28.
40. *Morale du joujou*, first published in 1853. I take the translation from *The Painter of Modern Life and Other Essays*, trans. Jonathan Mayne (London: Phaidon Press, 1964) and offer the original from Charles Baudelaire, *Œuvres complètes*, ed. Claude Pichois, vol. 1 (Paris: Gallimard, Bibliothèque de la Pléiade, 1975).
41. Baudelaire, *Painter*, 202, emphasis in both translation and original: "La plupart des marmots veulent surtout *voir l'âme*." *Œuvres complètes* vol. 1, 587.
42. Baudelaire, *Painter*, 203, emphasis in original: "*Mais où est l'âme?*" *Œuvres complètes* vol. 1, 587.
43. Baudelaire, *Painter*, 198. Original French: "Tous les enfants parlent à leurs joujoux." *Œuvres complètes* vol. 1, 582.
44. "Je ne veux pas parler de ces petites filles qui jouent à la madame . . . Les pauvres petites imitent leurs mamans : elles préludent déjà à leur immortelle puérilité future." *Œuvres complètes* vol. 1, 583. As Adam Rosenthal writes, because their play is "purely mimetic," the girls Baudelaire describes "remain with a restricted economy of maternal reserve," as a consequence of which Baudelaire rejects them. "The Gift of Memory in Baudelaire's 'Morale du joujou,'" *Nineteenth-Century French Studies* 43, no. 3–4 (2015): 129–143, 137.
45. *The Critical Review, or, Annals of Literature*, April 1792: 389–398, 397. A second scathing review would follow in June.

6

AUTOMATIC FOR ALL

Mary Shelley's Posthuman Passion

Kate Singer

After Percy Shelley's death, so the story goes, Mary Shelley kept herself alive scribbling for the popular press, writing short stories for annuals and magazines, which took time away from her more ambitious novel writing but helped to maintain a domestic life with her only surviving son.[1] Despite investigations into the stories' meditations on commodity culture, feminine beauty and embodiment, gender bending, and narrative experimentation, the tales have received little critical attention. "Valuable to biographers and critics for their personal content," they remain seen for the most part "wordy and pedestrian."[2] Mary Shelley's own comments about needing the "clear water" of a novel to "wash off the ~~dirt~~ mud of the magazines" and her thoughts that she was better at "biographical work than 'at romancing'" seem to support the idea that tales were her necessary servitude to bad romance: overblown love, thin causality, unreal plots.[3] Surges of tragic passion that power these tales especially can appear mechanical and passé, as if Shelley never read her mother's scathing critique of the culture of sensibility where overblown passion, taught to women by novels and conduct books, created them as creatures of unthinking sensation, intoxicated without a will of their own. Stories set in early modern Europe seem to attend to pre-Enlightenment passion with the ardor of the Victorian cult of death that bypasses Romantic critiques of sensibility almost entirely. As Shelley writes in the introduction to "Ferdinando Eboli," "[t]hose were more romantic days than these; for the revulsions occasioned by revolution or invasion were full of romance" (65).[4] However improbable these narrative strains—encased in the mechanical art of romantic reproduction, engendered by the story genre and its commodification in *The Keepsakes*, then sold to mass audiences hungry for easy love—this essay argues that they produce something more than instinctual, unthinking, automatic passion.

Shelley's tales of romance abound not with idle repetition but with transformation. A. A. Markley has reminded us that Shelley's gothic doubling, so familiar from *Frankenstein*, continues as a motif through her tales, both to alert us to the homosocial trajectories of its characters and to showcase Shakespearean cross-dressing that raises questions as to the performativity of gender and its limits during the period. But Shelley, ever the posthumanist, goes even further to create nonhuman transformations as well. In "Ferdinando Eboli," when the titular character confronts an imposter in vain, his anger turns him animal: "The rage of a wild beast newly chained was less than the tempest of indignation that now filled the heart of Ferdinand." (73). When the imposter, Ferdinand's older, natural brother, is confronted by his lover (stolen from his brother), "he turned—a stag at bay" (75). The lover, too, undergoes three changes during the story. Minutes before this confrontation, "she had been a young and thoughtless girl, docile as a child. . . . Now she felt as if she had suddenly grown old in wisdom" (75). When she escapes his tyranny, she dons the clothing of her mother's dead page, and when she seeks rest in a pirate cave, "[s]he seemed a being of another world; a seraph, all light and beauty; a Ganymede" (79). This rapid juxtaposition of human, animal, and superhuman characterizes many of Shelley's stories, as the largely automatic passion instigates shape-shifting across species and categories of human, animal, thing, superhuman, and cyborg. Shelley's play with the Great Chain of Being, what she called in her journal "the catalogue of beings," transmutes human love into a posthuman passion (471). As Allison Muri argues, automata form only one part of the history of cyborgs. Another important piece is the history of physiological mechanisms, including theories of atoms, energy, and material transfer, which enable us to see cyborgs as "human-machine-text moved by energy and controlled by a circuit of communication."[5] Shelley adds to this history when she posits passion as affect that moves among bodies and that has the power to alter them. It operates through an automaticity that is not so much a mechanical repetition but a recursive, material movement.

This essay attempts to uncover Shelley's subtle and pervasive thinking about how passion unsettles category distinctions throughout her stories, as her plots achieve finer tunings of affect as a mechanism of ontological change. While earlier stories including "Ferdinando Eboli," "The Mourner," "The Swiss Peasant," and "The Dream" explore a variety of body swapping, later stories such as "The Brother and Sister," "Transformation," and "Mortal Immortal" take a slower approach to portray passion's mechanism as a techno-chemical romance, which occurs through the intertwined discourses and materialities of passion, magic, and chemistry. Shelley's allusions to Cornelius Agrippa, in both *Frankenstein* and "Mortal Immortal,"

I want to argue, are no mere dalliance. Her science fiction mounts a syncretic investigation of materiality and affect, combining the electrochemistry of her day with Agrippa's accounts of alchemy's transformative abilities to explore a posthuman passion. This form of affective materiality, in turn, creates an autopoiesis that rewrites typical distinctions between human and nonhuman (even as it subverts Renaissance and Enlightenment humanities). If, as the character in "The Dream" asks, "Do not all things love," then Shelley pursues the Frankensteinian knowledge of how they do: in an automaticity that involutes human sentiment, winds, waters, and the spirit of the world in automatic movements (161). Shelley answers her husband's uncontrollable west wind with a cyborgian spirit of the world that neither humanizes technology nor conceptualizes the human as machine but entangles both.

TAINTED LOVE AND QUEER ANIMACIES

Pointing to the "great amount of technical artistry" in her stories," Markley calls our attention to their gothic doubling and their gender bending.[6] Sonia Hofkosh describes how the stories "enact this process" where women's bodies are both disfigured and then naturalized as different, like the commodified tales themselves.[7] These shiftings of bodies, selves, and their accoutrements, however, are part of a larger figural and material economy of bodily change that also mounts questions about ontological categories in excess of sex or subject-object distinctions. *Frankenstein*, of course, is renowned for its evocation of palimpsestic readings of the creature which mark him successively, though often not synchronically, as feminine, queer, working class, racial, nonhuman, and posthuman. There is similar precedence for such a shift from human to animal in Shelley's *The Last Man* (1826), where the apotheosis of Adrian and Lionel's first encounter is Lionel's transformation from savage wolf into a civilized human, then, at the novel's end, into the last man and the first posthuman.[8] In "Ferdinando Eboli" (1828), Shelley experiments with all these valences together, layered throughout the story, and in doing so articulates a larger question about how affect manifests in different species with differing levels of vitality and volition.

When Ferdinand returns home after having been captured during a secret mission in the Napoleonic Wars and the King decides in favor of the imposter, Ferdinand's emotions spike: "[i]t requires a strong imagination, and the experience of much misery, fully to enter into Ferdinand's feelings" (70). When Adalinda, his betrothed, is called upon to declare the true Ferdinand, and she too decides in favor of the imposter, her lover's judgment exacerbates his affect. His beastly rage surpasses the categories of emotion (such as sadness, anger, or joy) and resembles how Brian Massumi defines affect as intensity, "embodied in purely autonomic reactions," a "nonconscious" "temporal sink" "filled with motion, vibratory motion, resonation."[9] Intensity can be translated into language, static emotion, or directed activity but is inassimilable to it. Likewise, such affect exists before specific language or emotions are selected but also after it, as affect occurs in a feedback loop, much like Shelley's own narrative tactics which repeatedly revisit a character's nearly unaccountable feelings. Thus Ferdinand's "tumult of feelings" remains nebulous and largely inexpressible to the reader. Rather than effeminizing him or paralyzing him, they transform him into a wild beast, but one whose fortitude and force of spirit engage him to continue his pursuit of his rightful place—as he morphs from an aristocrat into a war-torn beggar, a pirate, a solider and, finally, something else entirely.

The generation of autonomous, excessive affect—in excess of emotion or expression—appears again when Adalinda, in "[e]xcess of fear," transforms from a child into an old wise person. A few sentences later Ludovico (the imposter brother) likewise becomes "a stag at bay" when "the energy and eloquence of truth [bears] down his artifice" (75). Shelley takes one more witty turn here: "Then it was her turn to quail; for the superior energy of a man gave him mastery," as Adalinda becomes a bird frightened by the charging, equine energy (75). She eventually becomes angelic and as superhuman is found by Ferdinand in a pirate's cove. Their reuniting this way secures an alliance of liminal beings on the edges of earthly civilization.

Such shape-shifting is at work in Shelley's other stories. In "The Evil Eye," Dimitri, having lost his family to rapacious invaders, becomes savage: "[h]is habits kept pace with his change of physiognomy . . . he had arrived at that worst state of ruffian feeling, the taking delight in blood" (101). In "Transformation," Guido's overwrought affect at his loss of patrimony and its love object leads him to the brink of a storm that manifests a dwarf who swaps bodies with him. In "The Swiss Peasant," Fanny's proper suitor, Henry, "spirited and as heedless as a chamois" with his carefree youth and later political ardor, resonates with Wordsworth's bounding roe in "Tintern Abbey" (141). The other suitor, Louis, from a ruined family, who becomes "the depositary of the many feelings" of Fanny, eventually comes to be on the other side of a political conflict so that "to appeal to him was to rouse a tiger from his lair," his passions as irruptive as the "fierce passions of the mob" (141, 148, 147). When Fanny finally pleads for Henry's safety, Louis becomes human again when "tide of warm, human, and overpowering emotion flowed into his soul." "The Brother and Sister" features men on both sides of a family feud with "superhuman spirit" (167, 170). During "Mortal Immortal's" declaration scene, when Bertha finally runs away from her suitor to Winzy, "she springs with a light fawn-like bound down the marble steps" while a page in her mother's retinue calls to her, "Back to your cage—hawks are abroad!" (223).

These comparisons between forms of life are peppered with the language of animacy. In "The Brother and Sister," after the brother challenges his family's enemy, fails, and is banished, his sister, Flora, is entrusted to his nemesis, Count Fabian. When he is thrown from a horse (incredibly near her villa), and Flora "beheld the inanimate body," she becomes his nurse, waiting on him with "that spirit of Christian humility and benevolence which animates a Sister of Charity" (179). Part of his recovery entails telling her "numerous *fabliaux*, *novella*, and romances," which remind her of the stories she had invented of the adventures she imagines her brother has had, "which animated her face as she listened" (180). If Flora reanimates the Count with her chaste ministrations, he does the same when he tells her stories that bring to life her own inventive tales. Storytelling and care become interwoven in a love song that multiplies affect's liveliness.

Once we understand the story to be about reanimation, we can read back several paragraphs to find another source of Flora's capacity, the nonhuman tutelage she receives in her rural retreat: "Blossoms and flowers, laughing plenty, graced the soil; and the trees, swelling with buds ready to expand into leaves, seemed to feel the life that *animated* their dark old boughs. Flora was enchanted; the country labors interested her, and the hoarded experience of old Sandra was a treasure-house of wisdom and amusement. . . . She learned the history of the bees, watched the habits of the birds, and inquired into the culture of plants" (178, emphasis mine). Just as the buds feel the life of the seeming old boughs, Flora's bird watching and floral curiosity enable a posthuman animacy that bespeaks the truth of her name. Unlike the botanists of the late eighteenth century, Flora is uninterested in categorizing the names of plants or entering into the discourse of Enlightenment sorting of species. Rather, she initiates a cross-species understanding of liveliness.

Mel Y. Chen has recently opened up the discussion of animacy to investigate how propositions about "liveliness, sentience, or human-ness" construct and justify hierarchies that subordinate those judged to be queer, animalistic, or racial.[10] Using the term as it was developed by linguists, Chen delineates an "animacy hierarchy" that places at the top adult man who is free to act as the most active cognitive agent (within language), followed by women, nonhuman animals, and even incorporeal forces (such as emotions or the wind).[11] Chen is particularly interested to "map the ways in which such a conceptual hierarchy cannot but fail, the ways in which it must continually interanimate in spite of its apparent fixedness."[12] It also, as Shelley does, ties animacy to "an ontology of affect: for animacy hierarchies are precisely about which things can or cannot affect—or be affected by—which other things within a scheme of possible action."[13]

Shelley's own catalogue of beings disrupts this hierarchy, as passion undoes rather than solidifies the division between rational, civilized man and the savage, instinctual animal. Shelley's unthinking humans-turned-animal resemble Descartes' "animal machine," whose complex internal structure, including the physiology of affect, operates without thought. Yet for Shelley, such unconsciousness is provocatively productive. Perhaps a second thing to note is how the concept of animacy—a linguistic sign of agential force, activity, or vitality—helps us move past the question of whether Shelley is simply comparing her characters to animals and superhumans. For Chen, language creates both a conceptual index of activity and also the activity that itself affects other beings, or the ability to affect change in themselves and others. Such shared activity, neither simply response nor reaction, can help us redefine life and vitality as the capacity for activity, a definition that sounds quite close to Massumi's affect as "incipient action and expression."[14] Life is affect, as the incipient or potentialized capacity for activity, for movement, not simply lively movement as "the cause of living phenomena," as John Abernathy thought, nor simply what William Lawrence supposed to be "the result of the working operations of the living body."[15] In ways we will see below, Shelley's notion of passion does not simply reproduce nineteenth-century vitalism or materialism, but it can help us to refigure life, organic and inorganic, as an infectious materiality with a capacity for unthinking change.

If for Massumi, such incipience occurs before expression or language, and for Chen, such potentiality resides within language's own activity, then in Shelley, I want to argue, the apotheosis of animacy occurs in affect's potential to instigate such a crossing over, to switch from one form of animacy, one species, to another: Flora as human and nonhuman. It might seem as though Shelley privileges a hierarchy when excess emotion transforms humans into savage animals subject to their unbridled passion and animal instinct. This reading would place her in conversation with Wollstonecraft when she compares women to slaves, animals, and blown flowers, having very little agency or animacy.[16] However, Shelley suggests that passion, as animating affect, allows us to see the posthuman always already there—and its great potential for psychological and ontological revolutions. Like Ferdinando and Adalinda, Fabian and Flora meet on equal ground only after both have been reanimated by nonhuman passion. Posthuman affect powers the "revulsions," "revolution or invasion" that are "full of romance."

Many of Shelley's more lyrical descriptions in "The Mourner" and "The Dream" evince such animacy, though not with animals but with winds and water.[17]

"The Mourner" opens on a "fair expanse of Virginia Water": "Earth, water, air, drink to overflowing, the radiance that streams from yonder well of light: the foliage of the trees seems dripping with the golden flood; while the lake, filled with no earthly dew, appears but an imbasining of the sun-tinctured atmosphere."[18] Shelley seems to recall Percy's own discussion of nonhuman affect in "On Love": "In the motion of the very leaves of spring, in the blue air, there is then found a secret correspondence with our heart . . . which by their inconceivable relation to something within the soul, awaken the spirits."[19] This secret correspondence is the animating movement, affect's capacity for activity, that corresponds to both human passion and to the overflowing radiance that plays on earth, water, and air at once. In "The Dream," Shelley is even more explicit: "Do not all things love?—the winds as they whisper to the rushing waters? The waters as they kiss the flowery banks, and speed to mingle with the sea? Heaven and earth are sustained by, live through, love" (161). The big love of romance, for Shelley, is a movement of one thing by another thing, one body affecting another body. Such affect not only reanimates but also initiates a process whereby all things can become another thing: human, animal, inorganic, or incorporeal. When these tales explore a new material affect that crosses and remakes ontological boundaries, they do so recursively (within tales and among them), further heightening the questions of affect's nonhuman automaticity.

MY CHEMICAL ROMANCE: SHELLEY'S MECHANISM OF NEW MATERIAL AFFECT

In later stories, Shelley develops her exploration of this materialist mechanism of passion. How is it, as the narrator says in "The Sisters Albano," that one "could die, but not cease to love" (55)? What kind of life would passion have outside the human will or a lively human body? As a materiality that can alter nature and one that depends upon repetition and automaticity, what does its unique technology consist of? To understand these speculative and realist possibilities of transformation through passion, we must turn to Shelley's syncretic examination of science, sentiment, and alchemy.

The science—or pseudo-science—endemic to *Frankenstein* has produced a body of scholarship attempting to negotiate Shelley's reading of the vitality debates as well as the most recent developments in physics and electrochemistry. Maurice Hindle and Laura Crouch have established Shelley's reading of Humphry Davy, who believed in electricity as the vital fluid moving within all living things.[20] Anne K. Mellor's seminal discussion of Shelley's feminist critique of science suggests that Shelley frowned upon the speculative hubris of Davy in favor of the more careful scientific practices of Erasmus Darwin.[21] Yet, both philosophers observe a universe where chemical mechanisms attend to copious change in form, aided by an animating fluid that appears in and as different forms of matter. Zak Sitter has recently argued that Shelley's *Frankenstein* critiques the materialist definitions of organic life expounded by Percy's one-time doctor, Lawrence, through the creature's inorganic matter, which dissolves the telos of organic structures in the novel: the human body, family, intention.[22] Shelley had also read Friedrich Schlegel's distinction between organic, which unfolds its structure from within, and the mechanical, which imparts form onto a lumpen mass.

Rather than coming down on either side of these debates, Shelley seems to play both ends against the middle by describing a materiality that itself moves through organic and inorganic, as both thing and process, regifting form not to the lumpen mass but to organic forms already unfolded. Stuart Curran gets us closer to this capacious figuration of affect when he argues that Shelley's limited use of scientific terminology in *Frankenstein* "turns scientific principle into a universal field of metaphor," one based on affinity: "the metaphorization of a universal electrochemical affinity between material bodies" regardless of gender.[23] As Curran hints, this affinity is chemical, alchemical, affective, and in some sense always metaphorical—something we can only see and experience through the turning of metaphors that link together different bodies.

Such materialism has, since Paul de Man, been tied to language's automaticity, yet within the period the empty repetitions of both language and matter were often joined to the discourse of sentiment and the sensible body.[24] The discourses of sympathy and sensibility produced a strand of thinking about affect as an unthinking physiological mechanism reproduced in the stock novels of sentiment. Alex Wetmore discusses how the man of feeling, and literary sentiment more generally, was tied to the idea of automata through "the mechanistic system of nerves and senses," which gave near universal access to moral sentiment, even as "it also threatens to destabilize categorical divisions."[25] Women were often condemned to being machine-like animals. As Wollstonecraft famously wrote, "they are treated as a kind of subordinate beings, and not as a part of the human species" (10). If women were animal-machines, so too was their feeling. Terry Castle reminds us of the female thermometer, a glass tube filled with hair and feminine essence worn to measure feeling, which conceptualized the human—especially feminine sentiment—as mechanical.[26] As Jessica Riskin argues, science and sensibility, whether human or animal, were expressly imbricated, as sensibility not only drew from science but was likewise constitutive of discoveries in chemistry and physics.[27]

What is perhaps even more interesting is how Shelley's own sentimental-scientific experiments sought to bind the conception of automatic affect and the mechanics of electromagnetic force together not through progressive science or advances in moral sentiment but through the flim-flam of alchemy. Perhaps because scholars have been so interested in unearthing Shelley's attention to "the cutting edge of the material sciences" and their radical modernity, there has been virtually no serious consideration of any veritable use of alchemy, despite Victor Frankenstein's infamous reading and in spite of Winzy's tutelage by Agrippa himself in "Mortal Immortal" (1833).[28] Shelley may have been influenced by her father's account of Agrippa (and other magicians) in *Lives of the Necromancers* (1834), where he tells the story of Agrippa's embarrassment at the hands of a clerk who raises a demon, yet unlike Godwin's project to lambaste magical credulity, she explores its sentimental science.[29]

Agrippa's philosophy—which now has a definition in the Stanford Encyclopedia of Philosophy[30]—is a particularly interesting mix of Neoplatonism (where each species has certain forms and properties), alchemy's concerns with transforming one species to another (lead into gold), and the isolation of the "spirit of the world" that enables these transformations. Concerning this "spirit" he writes, "[b]y this Spirit, therefore, every occult property is conveyed into herbs, stones, metals, and animals, *through the Sun*, Moon, Planets, and through Stars higher than the Planets" (70).[31] While we can access this spirit through study of nature and heavenly bodies or through the manipulations of sorcery, so can we access it through the passions—a topic which takes up the last section of the book. The passions, says Agrippa, "change the proper body with a sensible transmutation, by changing the accidents in the body, and by moving the spirit upward or downward, inward or outward, and by producing divers qualities in the members" (196). This "sensible transmutation" moves the spirit and thus various parts of the body, including its organs and indeed the body itself. Agrippa tells a story of a man dreaming of being a bull who woke up with horns: "and sometimes men's bodies are transformed, and transfigured, and also transported" (198). This most often occurs with the force of will, yet it also occurs through a melancholy disposition so great that it draws celestial spirits into human bodies.

Such a saturnine drawing in of the spirit of the world is an apt way to describe the gothic plot of "Transformation," where a broke and pining Guido arrives at the coast and almost seems to conjure the spirit of the dwarf who will help him. The story begins in terms of spirit, as Guido, speaking to readers, tells us "when any strange, supernatural, and necromantic adventure has occurred to a human being," he "feels at certain periods torn up as it were by an intellectual earthquake, and is forced to bear the inner depths of his spirit to another" (121). This language of ontological disruption and spirit continues when he describes himself as once an "imperious, haughty, tameless spirit" while Juliet is, of course, all that is gentle and pure, "enshrined in that celestial tenement" (122). Both possess some element of the spirit of the world, though Juliet more purely, while Guido remains in a constant tempest for the duration of the story.

It is this tameless temper that ironically turns him into something other than himself when it brings him to the shores of ontological change in the story's great act of supernatural conjuration. After having lost all his human worth—patrimony, money, land, pride, Juliet, and the faith of her father—he wanders the seashore, "a whirlwind of passion possessing and tearing my soul," when "[t]he clouds had strange fantastic shapes; and they changed, and mingled, and seemed to be driven about by a mighty spell" (126). Guido's turbulent, melancholic disposition is mimicked by the landscape in analogy but actually may conjure it into being, drawing the spirit of the world into manifestation or sharing affect with it. Shelley's "affinity among material bodies" occurs through shared affect, much as the "impersonal affect" that Jane Bennett describes in *Vibrant Matter*.[32] For Bennett, both organic and inorganic bodies have the Spinozan capacity for activity and responsiveness. Yet, unlike vitalists, her "impersonal affect or material vibrancy is not a spiritual supplement or 'life force' added to the matter said to house it."[33] "She equate[s] affect with materiality, rather than posit a separate force that can enter and animate a physical body."[34] Bennett aims to move past "passive, mechanistic, or divinely infused substance," with a notion of a "field of forces" that are both matter and energy at once. We can hear echoes of Shelley's own double-cross of the vitality debates, between John Abernathy's electric fluid and William Lawrence's mechanistic (if revolutionary) matter, as she expresses an interplay of material threads that affect and respond to each other, whose matter and motion, materiality and affective response, cannot be split.

When Guido meets the dwarf, this "magician" charms him into a trance, and his lack of volition combined with his responsivity puts him into relation with a slew of things also at the beck and call of the dwarf: "there was something fascinating in a being whose voice could govern earth, air, and sea" (128, 198). When he agrees to the dwarf's pact, Guido describes the act in terms of a will-less passion that sweeps him along: "Thus it is: place our bark in the current of the stream, and down, over fall and cataract it is hurried; give up our conduct to the wild torrent of passion, and we are away, we know not wither" (129). Here Guido's will is suspended by both his own passions and the power of the magician to summon and move the elements. Yet, perhaps most startlingly, neither Guido nor his passions are paralyzed; rather, they propel his shift. He echoes such sentiments with greater vehemence after he inhabits the dwarf's "shape of horror": "How does thou dart unknown pangs all through our feeble mechanism, now seeming to shiver us like broken glass, to nothingness—now giving us fresh strength, which can *do* nothing, and so torments us by a sensation, such as the strong man must feel who cannot break his fetters, though they bend his grasp" (130). In the first piece of this quote, before the dash, Guido spells out, through the language of the machine, an automatic, repetitive form of sensation and affect. "Unknown pangs," like Massumi's unfelt incipiencies, thrill through the body that channels them without the will of body or mind.

Those mechanical pangs alter that very body, here detrimentally, into "broken glass," and also shatter the thinking subject. From Guido's perspective at this point in the story, in the middle of his ontological and existential crisis, posthuman passion has destroyed his subjectivity and his body as well, as the passions fetter and "bend" his grasp. Yet, from the viewpoint of the story's end, his "person a little bent," he thinks of the magician as "a good rather than an evil spirit," so that such dissolution of the Enlightenment subject and stable body becomes his salvation (135). Guido's passionate fetters bend the masculine body and its patrimony when he becomes a "fonder and more faithful husband" (135). This is not a return to heteronormative patriarchy, but rather the spirit of the world has altered Guido's animacy incrementally, through his dwarf being, toward a slightly larger capacity for activity, moving and moved by all sorts of things. His passions orient him, as Sara Ahmed would say, to the roiling clouds, to his wife, and to the now good spirit of the dwarf.[35] Shelley's characteristic irony reveals how the mechanical qualities of impersonal affect and responsive matter do not produce humanistic agencies in dwarves and clouds. Rather, passion's agential force is a kind of unthinking response, not sheer repetition but a posthuman recursion that enacts change without self-conscious will.

THE WIND BENEATH MY WINGS: ALCHEMICAL SPIRITS AND TECHNOLOGICAL MEDIUMS

Not only does Shelley investigate the contretemps of autonomy and agency within affect, pointing us to the unexpectedly productive field of mechanistic force, but she also considers the role of technology in affect, or affect as a hybrid *techne*. This line of thinking is most evident when she ties both passion and magic to the science of chemistry, as she does with the potion that Winzy drinks to activate his own transformation in "Mortal Immortal." Winzy becomes the last apprentice standing after Agrippa's embarrassment. He allegedly sticks around for the gold the old man gives him, yet even this payment is not enough to gain Bertha, a fact that suggests that Winzy stays because he is drawn, with alchemical force, to the technology of affect. He repeatedly notes the chemical nature of Agrippa's preparations, "feeding his furnaces and watching his chemical preparations" as well as tending the "changes of his unintelligible medicines" and the "progress of his alembics" (221). This activity unsurprisingly parallels Bertha's own "haughty spirit" that "fired" at Winzy's neglect, such that his attention to Agrippa's scientific inventions, his nonhuman infatuation (however compelled), has an inverse relation to his intentions for his human passion (221). We might easily read Winzy's unwilled, continued subservience to Agrippa as an allegory for Shelley's own: she too needs funds for her loved one, and serves the defunct and derided Agrippa as her master, whose supernatural pyrotechnics are both false and alluring to mass audiences. Yet, another explanation may lie more clearly within the figural register of the story: like Winzy, Shelley continues to attend to Agrippa because her passion is already entangled in unfinished chemical experiments.

It is no surprise, then, that Winzy mistakes Agrippa's career-ending philter as a potion to cure love, when, as Agrippa later tells him, it is a "cure for love and for all things" (224). Winzy thinks he is ingesting something to stop the motions of love, his matter and energy reaching out to Bertha's own fiery whirlwind. Yet, the medial spirits he adds to his body are a potion that will alter his frame and soul: "[t]he bounding elasticity of the one—the buoyant lightness of another" (226). We could call the "glory and bliss" he experiences just after drinking an intoxicating *jouissance*, even largely homosocial in its Frankensteinian laboratory ardor (223). Agrippa argues that such transmutations occur through "the spirit of the world," "the *medium*, whereby Celestial Souls are joined to gross bodies, and bestow upon them wonderful gifts" (70). The spirit of the world is not made up of any of the elements (air, water, fire, earth) but a first thing that is before or beside them. This medium is thing and technology much like language itself: both material and process, tool and technique, a formless form that is "as much force as entity, as much energy as matter," as Bennett writes (20). Winzy's transformation, therefore, is nothing less than the alteration of body and spirit at once, not what Percy in *A Defence of Poetry* called "adding spirit to sense," or simply creating a new animal body, but instead altering the human into a cyborg, a "hybrid of machine and organism."[36] Donna Haraway identifies this hybrid in the nineteenth century as a "dialectical progeny called spirit or history" that settled the dualism between materialism and idealism.[37] Although Haraway argues that pre-cybernetic machines (such as automata) "were not self-moving, self-designing, autonomous," in fact, "the progeny called spirit" in Shelley's stories offers an affective materiality as contagious self-altering technology—propelling new forms of life and love.[38]

Winzy notes that he and Bertha do not—cannot—have children, as if his transformation suspends heteronormative passion that leads to reproduction of the species. Yet, their dual transformation (for Bertha's "vivacious spirit" becomes an "ill-temper") does not signal a cryogenic suspension, or what Robert Mitchell has described as the "suspended animation" of life without movement (227).[39] He is certainly not a human who can feel the categorically normative emotions of "ambition nor avarice." The "ardent love that gnaws at [his] mind can never be returned," because his love is no longer human but something cyborgian. Not so much an immortal human, Winzy becomes a spirit of posthuman passion, a roving cyborg, an experimental being of the world.

In the final lines, he voids his Enlightenment subject and body in a revolutionary suicide, which both William Godwin and Wollstonecraft favored, and augurs a new existence: "I yield this body, too tenacious a cage for a soul which thirsts for freedom, to the destructive elements of air and water—or, if I survive, . . . I shall adopt more resolute means, and, by scattering and annihilating the atoms that compose my frame, set at liberty the life imprisoned within . . . to a sphere more congenial to its immortal essence" (230). Winzy dissolves his body into the spirit that moves among air and water (two elements Agrippa thought able to communicate with multiple bodies and elements), one that manifests beyond the imprisoning chains of temporally and spatially located bodies or animacies. Within this passive decay and destruction, he shirks even the atoms that would (in a Lucretian universe) solidify the body as matter, and he transforms into an essence with a capacity to move amid all spheres, not simply the dim earth. The technology of spirit of the world, as medium, highlights affective materiality as both the instrument of movement and the thing that moves, the communication and the channel. Bodies become mediums of affective transmission even as they share and emit autonomous material communications. We could call this participatory responsiveness a mechanistic reaction, or we could think of it as a cyborg passion—a material fluidity of recursive reactions both before and after human and nonhuman motions.[40]

PASSION'S POSTHUMAN RECURSION

What kind of posthuman is such a cyborg essence—and what kind of cyborgian existence is posthuman passion? We could easily read Shelley invoking a critique of Winzy as an inept and erstwhile transhuman whose attempt at self-perfection through technology serves as a critique of the anthropocentric, rational philosophies deeded by empiricists, Kant, and Godwin's own notions of perfectibility.

In the name of progress, both rational philosophy and transhumanism tend to shirk embodiment, as Cary Wolfe argues, secretly underwriting the human/animal and human/nonhuman divides.[41] For Wolfe, truly revolutionary posthumanism "opposes the fantasies of disembodiment and autonomy" and embraces myriad forms of embodiment. That is, the imbrication of the human within "technical, medical, informatics, and economic networks is increasingly impossible to ignore."[42] As we have seen, Shelley already depicts these multiple forms of entangled embodiment. Agrippa's "unintelligible medicines" intimate both early modern and Romantic experiments in natural science. The dwarf's storm-born treasure speaks to the dangerous and miraculous economic networks of mercantile speculation and stocks. "Ferdinando Eboli" sketches the ever-present military tools of espionage, travel, and death but also the technologies of wartime information. When Shelley's stories demonstrate how bodies themselves yield to posthuman passion, they generate an affect that recursively alters its nature each time: into animal rage, passion's whirlwind, Agrippan spirit, chemical invention, and the most pliable technology of all, a materiality conversant with the world.

Yet Shelley's most valuable discovery may be in a form and method of passion that becomes a means of opening the human to the nonhuman, as its automaticity moves beyond merely human-centered thought. Passion's recursive tendency toward automatic change as well as its imbrication of humanity, animality, science, and technology together model an approach akin to what Wolfe (following Niklas Luhmann) terms "openness from closure."[43] This second-order systems theory describes a system that in the act of closing itself, opens itself back up to its environment and the other systems surrounding it, recreating itself anew. Eve Sedgwick terms such a moment the "cybernetic fold," when systems thinking develops as cybernetics is on the rise, but before computation.[44] This cybernetic openness from closure is formed specifically through autopoiesis, a self-altering system of thought. For Wolfe, autopoiesis is a means for humans to get outside of anthropocentric thought; in Shelley, passion, not thought, propels us to understand and to be something not quite ourselves.

The autopoiesis of passion is not simply repetitive, but in reiterating its rages and tumults, its recursive motions produce all kinds of beings. It imbricates and alters them, creating a self-reproducing system whose supposed sameness incorporates things outside itself that then transform its trajectories and ontologies. We have, I think, already seen such a cybernetic loop in "Ferdinando Eboli's" recursive shape-shifting, as each return of passion forges a new change in bodily form, which alters the system of posthumanity by incorporating new ontologies. This automaticity actually moves us out of the human/animal or human/nonhuman divide based on rational perfectibility. A nonhumanistic form of thought and being arises through passion's self-maintaining, self-altering system of material movement and recursive change. It bespeaks an impossible animation—a sentience built on passion's response, contingency, and ontological fluidity. Reanimation then, for Shelley, is neither a Frankensteinian bringing the dead (or mechanical) back to life nor the positing of a life beyond the human. Instead, it recreates being from passion's loop-de-loops to change the system of humanistic thought altogether. Such a model appears perhaps most clearly and meta-poetically in Shelley's two reanimation stories: "Valerius: The Reanimated Roman" (c. 1819) and "Roger Dodsworth: The Reanimated Englishman" (c. 1826). The posthumous existence of these two stories (with "Valerius" never brought to life in publication) will close the loop that opens Shelley's romances to the revolution of posthuman passion.

The narrator of the later story, a speculation on Dodsworth's reaction to the progress of Western civilization, ponders "how such and such heroes of antiquity would act, if they were reborn in these times" (48). As Charles E. Robinson notes, Shelley connected the two stories, as the discursive "Roger Dodsworth" sends us back to the sentimental narrative of the Roman "Valerius," so that we may open ourselves to even more radical reanimations, not of heroes, but a posthumanity that fuses nature, culture, and the supernatural (397). The earlier story records Valerius's first-hand account of his grand disappointment in Rome's ruins. When he awakens, Valerius is shown Rome by a Catholic priest who obsessively dwells on its decay and deterioration. His masculine perspective only understands history in terms of the death of republican humanism and the tumescent imperial monuments that speak to a life that disregards much of humanity. However, when he meets a young Scotswoman, Isabell Harley, their affective transference, aided and abetted by nature that also shares affect with them, inaugurates both his posthuman existence and her own.

The abrupt switch to Isabell's adjacent, sentimental narration supplies this second resonance to Rome's ruins, and this autopoiesis of affect disrupts the political systems built and cried over by men as it suspends the *agon* of death and life in favor of another transformation. As Isabell reports, "I wished to interest the feelings of Valerius and not so much to shew him all the remains of his country as to awaken in him by their sight a sentiment that he was still in some degree linked to the world" (342). The enlivening that almost occurs when they visit the "wandering Tiber" finally happens one night when they visit the Pantheon and see it bathed in moonlight: "[t]he spirit of beauty seemed to shed her rays over her favoured offspring and to penetrate every thing—even the human mind—with a soft, still bring glory" (342). This affect of the moon—its material rays penetrating even stone and marble—"may inspire virtue and love, but the feeling is far too intense for expression" (342). Rather than simply an intersubjective, trans-historical human sympathy between two people, the Roman is reanimated, not in body but in spirit, by a literal transmission of material affect. Such a vibrant matter is necessarily shared by the moon and the Pantheon.

Ironically, Isabell's attraction to the cross attached to the Pantheon disrupts their moment of exchange. We might say her Christianity, and its narrative of imperial war, tropes her out of this affective assemblage and back into a coercive, Christian nineteenth-century politics. Yet it perhaps more radically enacts an automaticity of affect—the resonating affects of the passion of Christ's own continual resurrections. Passion's catholic attachments figure a systemic change constantly on the rise—new systems of humans and their technologies, from Roman Republic to Roman Empire to Christian New Testament, to holy wars and back again.

After this moment, Isabell has an "inexplicable" feeling "that my companion was not a being of the earth . . . [h]is semblance was that of life, yet he belonged to the dead" (343). She is something like repulsed by him, followed by a spontaneous striving to repay this uneasiness with interest and intellectual sympathy, even cognizant of "the earthly barrier there seemed placed between us" (344). Once again, Isabell charts a resonance that alters the system of their relationship, and then resumes an automaticity of affect that speaks to something beyond the human that modifies her as well, down to the very pulses and repulses of her body. This posthuman affective exchange sends Valerius to England to learn its strangeness and to impart his own. Although Isabell attempts to integrate Valerius into lively, human sentiment, their affect, shared with the monuments and the landscape, alters them both alongside the terms and systems of their lives.

Shelley finds openness from the seeming closures of romance's sentiment, a technique that breaks down the barriers between fiction and discourse, human and nonhuman, animal and human, organic and inorganic, mechanistic and animated. Her theory of affect, rather than signaling either feminine excess or the mechanicity of sentiment in the nineteenth century, offers a means of leaving domineering human agencies behind to emerge into a future fluctuating with the spirits of the world. Shelley's passion both forms and powers the system of autopoiesis that might, through passion's recursions, propel us beyond human thinking and feeling.

NOTES

1. See Sonia Hofkosh, "Disfiguring Economies: Mary Shelley's Short Stories" in *The Other Mary Shelley: Beyond Frankenstein*, eds. Audrey A. Fisch, Anne K. Mellor, and Esther H. Schor (New York: Oxford University Press, 1993), 204–219; A. A. Markley, "Mary Shelley's 'New Gothic': Character Doubling and Social Critique in the Short Fiction," *Gothic Studies* 3.1 (2001): 15–23; A. A. Markley, "'The Truth in Masquerade': Cross-Dressing and

Disguise in Mary Shelley's Short Stories" in *Mary Shelley's Fictions: From Frankenstein to Falkner*, ed. Michael Eberle-Sinatra and Nora Crook (New York: St. Martin's Press, 2000), 109–126; and Elena Anastaski, "The Trials and Tribulations of the *revenants*: Narrative Technique and the Fragmented Hero in Mary Shelley and Théophile Gautier," *Connotations* 16.1–3 (2006/7): 26–46.

2. Miranda Seymour, *Mary Shelley* (New York: Grove Press, 2000), 338. Even Tilottama Rajan's thick explication of the abyss of affect in Shelley's later novels often relies on Shelley's psychoanalytic and biographical resistances to Percy Shelley, William Godwin, and Lord Byron to analyze affect and character. Rajan, "A Peculiar Community: Mary Shelley, Godwin, and the Abyss of Emotion," in *Romanticism and Emotion*, ed. Richard Sha and Joel Faflak (Cambridge University Press, 2014), 145–170.
3. MWS to Leigh Hunt, *The Letters of Mary Wollstonecraft Shelley*, ed. Betty T. Bennett (Baltimore: Johns Hopkins University Press, 1980), 1:412; MWS to Edward Moxon, quoted in *The Journals of Mary Shelley*, ed. Paula R. Feldman and Diana Scott-Kilvert (New York: Oxford University Press, 1987), 2:533n2.
4. Mary Shelley, *Mary Shelley: Collected Tales and Stories*, ed. Charles E. Robinson (Baltimore: Johns Hopkins University Press, 1976).
5. Allison Muri, *The Enlightenment Cyborg: A History of Communications and Control in the Human Machine, 1660–1830* (Toronto: University of Toronto Press, 2007), 115.
6. Markley, *Mary Shelley's Fictions*, 109, 119.
7. Hofkosh, 205.
8. See Chris Washington's chapter on Mary Shelley in *Romantic Revelations: Visions of Post-Apocalyptic Life and Hope in the Anthropocene* (Toronto: University of Toronto Press, 2019).
9. Brian Massumi, *Parables for the Virtual: Movement, Affect, Sensation* (Durham, Duke University Press, 2002), 24, 25, 26.
10. Mel Y. Chen, *Animacies* (Durham: Duke University Press, 2012), 24.
11. Chen, 26–27.
12. Chen, 30.
13. Chen, 30.
14. Massumi, 30.
15. Sharon Ruston, *Shelley's Vitality* (New York: Palgrave Macmillan, 2005), 13.
16. Mary Wollstonecraft, *A Vindication of the Rights of Woman*, ed. Deirdre Shauna Lynch (New York: Norton, 2009), 10.
17. Chen, 27.
18. Shelley, *Stories*, 81.
19. Percy Shelley, *The Collected Poems of Percy Bysshe Shelley*, ed. Donald H. Reiman and Neil Fraistat (New York: W. W. Norton & Co., 2002), 504.
20. Maurice Hindle, "Vital Matters: Mary Shelley's *Frankenstein* and Romantic Science," *Critical Survey* 2, no. 1 (1990): 29–35, 34.
21. Anne K. Mellor, *Mary Shelley, Her Life, Her Fiction, Her Monsters* (New York: Routledge, 1989).
22. Zak Sitter, "Inorganic Intentions: Organizing Form from *Frankenstein* to de Man," *Literature Compass* (13 Oct. 2016): 655–662.
23. Stuart Curran, "The Scientific Grounding of *Frankenstein*" *Mary vs. Mary*, ed. Lilla Maria Crisafulli (Naples: Liguori Editore, 2001), 283–292, 290, 291.
24. Paul de Man, *Aesthetic Ideology* (Minneapolis: University of Minnesota Press, 1984).
25. Alex Wetmore, *Men of Feeling in Eighteenth-Century Literature: Touching Fiction* (New York: Palgrave Macmillan, 2013), 71.
26. Terry Castle, *The Female Thermometer: Eighteenth-Century Culture and the Invention of the Uncanny* (New York: Oxford University Press, 1995), 21–43.

27. Jessica Riskin, *Science in the Age of Sensibility: The Sentimental Empiricists of the French Enlightenment* (Chicago: Chicago University Press, 2002).
28. In this approach, Romanticists may have placed themselves in the role of Victor's father, condemning attention to early modern science and alchemy as both passé and hubristic. But Shelley may be not so much critical of how Victor develops a syncretic science but of what he does with it—how he abjects and erases passion from a project that intertwines passion and science from the start. I am indebted to Lily Gurton-Wachter for her essay arguing for a longer history of "occult sympathy." Gurton-Wachter, "Sympathy Between Disciplines," *Literature Compass* (13 Feb. 2018): n.p.
29. While there is no evidence that Shelley read Agrippa, she certainly would have been familiar with him from Godwin's description in *The Lives of the Necromancers* as well as the alchemy in Godwin's *St. Leon*, whose titular character pursues the elixir of life only to become a wanderer like Winzy.
30. https://plato.stanford.edu/entries/agrippa-nettesheim.
31. Henry Cornelius Agrippa, *Three Books of Occult Philosophy or Magic*, ed. Willis F. Whitehead (Chicago: Hahn & Whitehead, 1898; rept. New York: AMS Press).
32. Jane Bennett, *Vibrant Matter: A Political Ecology of Things* (Durham Duke University Press, 2009), xii.
33. Bennett, xiii.
34. Bennett, xiii.
35. Sara Ahmed, *Queer Phenomenology: Orientations, Objects, Others* (Durham: Duke University Press, 2006).
36. Reiman and Fraistat, 528; Donna Haraway, "A Cyborg Manifesto: Science, Technology and Socialist-Feminism in the late Twentieth Century," in *Simians, Cyborgs and Women: The Reinvention of Nature* (New York: Routledge, 1991), 149–181, 149.
37. Haraway, 149.
38. Haraway, 149.
39. Robert Mitchell, "Suspended Animation, Slow Time, and the Poetics of Trance," *PMLA* 126, no. 1 (2011): 107–122, 110.
40. Richard Sha has recently written on affect's force as having elements of both human will and physical or chemical autonomy. My argument follows this doubleness yet leans toward an underlying materiality whose recursions build both volition and autonomy. Sha, "The Motion behind Romantic Emotion," in *Romanticism and the Emotions*, ed. Joel Faflak and Richard Sha (Cambridge: Cambridge University Press, 2014): 19–47, 40–41.
41. Cary Wolfe, *What Is Posthumanism* (Minneapolis: University of Minnesota Press, 2010).
42. Wolfe writes, xv, that posthumanism occurs "before in the sense that it names the embodiment and embeddedness of the human being in not just its biological but also its technological world, the prosthetic coevolution of the human animal with the technicity of tools and external archival mechanisms. . . . But it comes after in the sense that posthumanism names a historical moment in which the decentering of the human by its imbrication in technical, medical, informatics, and economic networks is increasingly impossible to ignore."
43. Wolfe, xvi.
44. While Sedgwick places this moment around the late 1940s–1960s, Wolfe might say that this kind of folding was always available as humans themselves developed with technology. In this vein, it would be tempting to consider how Shelley's association with Dionysus Lardner, editor of various encyclopedias as well as a natural scientist, might have introduced her to ideas about computation, as Lardner helped to write about the computational problems that influenced Charles Babbage's development of the difference engine. Eve Kosofsky Sedgwick & Adam Frank, "Shame in the Cybernetic Fold: Reading Sylvan Tomkins" *Critical Inquiry* 21, no. 2 (1995): 496–522.

7

"A LITTLE EARTHLY IDOL TO CONTRACT YOUR IDEAS"

Global Hermeneutics in Phebe Gibbes's *Zoriada, or, Village Annals* (1786)

Kathryn Freeman

I

In her triple-decker novel, *Zoriada, or Village Annals*, Phebe Gibbes creates a dynamic twist to the British oriental tale, a popular eighteenth-century genre taking its readers to exotic locales through the traditional subjectivity of an Englishman traveling to the east.[1] Gibbes upends the genre through incongruity, reversing direction and gender for her protagonist by setting the novel in a rural English village among whose provincial inhabitants a young Indian woman appears.[2] Armed with tools of rational investigation available to the inquiring Enlightenment mind, villagers and reader alike repeatedly attempt to discover the identity of the stranger, known only through her pseudonym, Zoriada.[3] The villagers represent a spectrum of hermeneutical strategies ranging from an unschooled caretaker's gossip reliant on a culture of objectification, to a rapacious man's aggression in trying to determine whether Zoriada is witch or angel, to a physician's futile scientific deduction in seeking the source of her suffering in order to heal her. In spite of the villagers operating from a clear division of social classes, their dualistic approach to solving the mystery of Zoriada's identity is the common denominator that contrasts with Zoriada's own resistance to the collective binaries they project upon her.

Leveling the investigative urges of the villagers, from the unschooled to the highly educated, Gibbes thus participates in and complicates the late eighteenth-century revolt against the Enlightenment's system of binaries underlying its celebration of deductive reasoning as a masculinist enterprise. Gibbes withholds not only her protagonist's identity as knowable through binaries including east/west, subject/object, masculine/feminine, and highbrow/illiterate; more importantly, she withholds her protagonist's subjectivity: the distinction is a new one in the late eighteenth century, distinguishing traditional notions of identity formed through ancestry, culture, and relationships from the deeper explorations of selfhood emerging through the ontological and epistemological breakthroughs that came to be associated with Romanticism in its revolt against western materialism.[4] At various points in the developing plot, Gibbes calls metatextual attention to the reader's complicit engagement in imposing a marionette-like shell of objectification on her heroine; her refusal to disclose her identity ultimately plays upon the reader's need to project the same system of binaries that informs the flawed judgement of the characters in the novel.

Through Zoriada's emerging subjectivity, Gibbes simultaneously transforms both the traditionally objectified, exotic female of the oriental tale and the generic heroine of the sentimental novel's marriage plot.[5] Though the novel resolves questions of her identity by Volume III, Zoriada's subjectivity remains her own to withhold, defined by neither her ancestry nor her role in the novel's interconnected threads. By refusing to endow her third person narrator with omniscience, and by having Zoriada refuse the probing demands for her identity, Gibbes ends the novel with villagers and reader no closer to Zoriada's subjectivity though perhaps too distracted by the busy plot to realize the novel's distinction between identity revealed and subjectivity withheld.

Gibbes's challenge to narrative commonplaces founded on Enlightenment dualism begins with the novel's title. Its either/or construction, *Zoriada, or, Village Annals,* announces the collision of two spheres traditionally separated not only through geography and culture but also epistemology. In terms of subject/object duality, the title appears to give the reader a choice of which lens s/he will use to view the events of the novel: Zoriada's subjectivity or the village "annals," a term implying a discursive history composed of the villagers' collected stories. The novel itself deconstructs the simplicity of the title's apparent choice on both sides of the comma's divide: to the left of the comma, that Zoriada is not the name of the stranger suggests one cannot enter the story through her subjectivity to which even the third person narrator is denied access; to the right of the comma, the villagers' stories, a collection of misguided or sometimes nefarious assumptions, are hardly the codified history the term "annals" suggests.

The two sides of the title unspool through the novel's double trajectory: underneath the convoluted and unlikely plot, spanning generations and zigzagging across the Atlantic, Gibbes steadily resists the objectification of her protagonist by characters who, projecting identity onto Zoriada, view her with a potent ambivalence of attraction and repulsion. By Volume III, however, these assumptions meet their ultimate challenge when many of the villagers discover their own stories are bound inextricably with hers. Zoriada is not only of mixed English and Indian parentage; to the surprise of the characters and reader, Gibbes belatedly integrates Zoriada's bloodlines and culture with those of the villagers: not revealed until the end of the novel, Zoriada's mother was a Muslim married to the betrayed brother of Crosby, one of the villagers introduced early in the novel; the man Zoriada chooses to marry at the end of the novel, Edmund Mims, is the son of the Anglo-Indian sea captain who had rescued her after her family was murdered and their home in India burned. Captain Mims, in turn, is the son of the village doctor, Withers, and his wife. The "village annals," stories within stories including betrayals, families torn apart, and lost children, thus form a lens to view the political, religious, social, and gender issues at stake in the characters' projection of identity onto Zoriada. The sophisticated plot structure sustained over the novel's three volumes is paradoxical, simultaneously embracing and evading the hermeneutical process of reading. Gibbes achieves this narrative deconstruction through repeated gestures to the novel's metanarrative commentary: in the quest to find meaning in the text, the reader is complicit in the villagers' projecting identity onto Zoriada, an empty center in a novel whose multiple subplots, filled with violence, intrigue, and a complicated network of relationships, appear designed to overwhelm the reader.

Gibbes deepens the novel's metatextual irony through repeated references to the Arabic language, pairing her ineffable heroine with an equally ineffable Arabic manuscript that one of the villagers, the rapacious Parson Swinborne, steals from Zoriada's room, hoping to deduce her identity from it. The reader implicitly colludes with the nefarious Swinborne, peering over his shoulder to identify the objects in her room. That the manuscript remains untranslated to the end of the novel, however, defies the villagers' and reader's desire to do no less than include it in a monolithic western history that the phrase "village annals" suggests. The protagonist's chosen pseudonym, "Zoriada," an Arabic name with a long literary history, further underscores Gibbes's complication of the hegemony of western hermeneutics. Commonly translated as "enchanting" or "dawn," Zoriada is not only the name of a saint who converted from Islam in twelfth-century Spain, but one that Gibbes probably knew from *Don Quixote* as the Muslim woman Cervantes associates with the Virgin Mary; the irony of the name vis-à-vis both figures of conversion from Islam to Christianity emerges gradually in the novel, as Gibbes presents a contrasting erosion of Christian hegemony culminating in the final revelation of Zoriada's mixed Christian and Muslim parentage.[6] Taking the reader beyond the collection of projections onto and belatedly discovered connections to Zoriada, the novel thereby challenges the dualistic fallacy underlying the objectification of Zoriada. Remaining a cypher until the novel's end, she and her manuscript embody a nondualism that eludes western binaries and, ultimately, the failure of western hermeneutics to crack the code of subjectivity.

II

From the novel's opening, Gibbes upends narrative commonplaces of the eighteenth-century novel founded on western binaries. A double objectification introduces Zoriada in Volume I through the conversation between characters representing twin poles of the novel's clashing social classes—the old physician, Doctor Withers, and the housekeeper of Heath-farm, Mrs. Leland, who has called the doctor to attend what she has assumed to be a "dying young lady" (I, 1). Embodying Della Cruscan emotionalism, Mrs. Leland is "bathed in tears" over what she assumes to be the imminent death of this young woman, done in by "close study of learning"; foreseeing the stranger's tragic end "as clear and certain, as if she had been shot through the head," the uneducated Mrs. Leland is the object of Gibbes's satire of the common eighteenth-century superstition that educated women subject themselves to untimely deaths (I, 1–2).[7] A foil to the educated Doctor Withers, Mrs. Leland is a gossip trafficking in the sentimental clichés of her time, speculating that the stranger may have been "stolen from her rich relations" in India or perhaps has never had "any earthly relations at all" (I, 22). Though we later learn that Captain Mims rescued Zoriada after the murder of her family and destruction of her home, at this early point we are given no narrative authority to debunk speculation and gossip. To Mrs. Leland's frustration, Mrs. Quinbrook, the captain's friend who has put the young woman in Mrs. Leland's care, has no information about her. That Mrs. Quinbrook only tells her Mims brought the stranger from India leads Mrs. Leland to suspect Mims "was dying in love with her" though "so much older than herself," misinformation remaining uncorrected until the end of the novel when, along with the villagers, the reader discovers Mims to be the father of Edmund, the man Zoriada chooses to marry (I, 22).

Gibbes uses Mrs. Leland's lowbrow gossip as a foil not to highbrow snobbery as one might expect in traditional eighteenth-century satire, but rather as a foil to the failure of Dr. Withers's scientific approach; expert in deductive reasoning, he represents the other end of the hermeneutical spectrum from Mrs. Leland, his name resonating with irony: the Age of Enlightenment "withers" by the late eighteenth century. More than a mere caricature of the man of science, however, Doctor Withers asks Mrs. Leland to let the stranger tell "her story in her own words," suggesting his sophistication in knowing what he does not know; his request also suggests Gibbes's metanarrative commentary since this desire for information about the stranger drives the reader's interest in the plot to the end of Volume III, the villagers fabricating her story based on the fantasies they project upon her.

Trained in deductive reasoning, Doctor Withers tries but fails to place into the western paradigm not only the stranger herself but the "striking tokens" in her room that defy facile categorization within the system of binaries forming the basis for his diagnosis. Withers does not see the common pattern among the seemingly incongruous objects: those involving sound include a lute, a musical instrument evolved from a common ancestor to both eastern and western music, and a "repeater," an intricate part in an eighteenth-century timepiece chiming the hour; taken together they suggest the stranger's cultural fluidity and paradoxical interest in sound, both musical and mechanical.[8] The most enigmatic and binary-defying object Withers finds in her room is an "armillary sphere"; this astronomical model connects Zoriada to an east/west nondualism, for it was simultaneously created in ancient Greece and China, later adopted in both the Islamic world and Medieval Europe (I, 3).[9] Unable to deduce her identity through these signifiers, Withers turns to her art but neither describes nor therefore interprets it; he concludes that the drawings are "the labors (it was evident) of her pencil," the parenthetical aside a limp attempt at rational deduction (I, 3). Withers determines, nevertheless, that she is a "superior order of beings"—superior, it would seem, not only to the quaint Mrs. Leland but even to himself (I, 3); though he easily determines the young woman is not dying but has merely swooned, the cause of her deep misery is inaccessible to him. He nevertheless warns Zoriada that her constitution is too delicate to bear "violent shocks"; yet the emerging details of her melodramatic story prove him wrong: she has undergone and will continue from this point forward to undergo shock after shock (I, 8). At this early stage of her story, she assures Dr. Withers, "I am accustomed . . . to such temporary suspensions of sense and motion, yet apprehend no fatal consequence" (I, 8). Gibbes thus distinguishes between the Della Cruscanism of Mrs. Leland and Zoriada's swooning, a manifestation of selfhood that eludes the doctor's training in rational deduction.

Attempting to convince Zoriada that her swoon should not be taken lightly, Doctor Withers compares her body to "a human machine" whose "springs . . . cannot be so suddenly let down without sustaining great injury"; when Zoriada rejects his scientific account of the body, saying, "I am no friend to medicines," Doctor Withers responds that the pulse is the "little pendulum of life," a phrase he may have gotten from Galileo's "pendulum clock" to regulate the pulse (I, 8–9).[10] If so, the allusion is particularly ironic, since Doctor Withers attributes the metaphor to "an English writer" rather than Galileo, the figure established as the creator of the scientific method (I, 9).[11] By invoking the "pendulum clock" metaphor, Doctor Withers imposes on Zoriada his ontological materialism, thereby connecting her to the objects he has scrutinized in her room.

Neither caricatured nor idealized, Doctor Withers is nevertheless humanized by his concern for Zoriada; he attempts to restore her health through the Romantic prescription of walks through "artless" nature, all elements of which "appear'd conscious of their peculiarly happy condition" (I, 34–35). Though she feels "insensibly harmonised" when Doctor Withers attempts to "regulate her feelings" by showing her his temple of resignation, Zoriada bursts into tears, saying her mind is incapable of learning resignation (I, 35, 45). However well-intentioned, Withers makes matters worse for the grieving Zoriada by offering as a parable of resignation how he and Mrs. Withers have coped with their own tragedy, the early losses of their young son and infant daughter. The story takes on particular irony since, having just warned Zoriada that she is too delicate to bear violent shocks, Withers tells of the accidental death of his baby daughter thirty-eight years before due, indirectly, to his assumption that a cold bath would "confirm the promise of a good constitution, and brace the tender nerves of our little girl" (I, 43–44). He explains that a woman who had been left alone to bathe the baby for the first time wrapped her too tightly in a blanket following the cold bath, suffocating her. Perhaps defensively, Withers says he had ordered the bath "in the fullness of human wisdom" (I, 44). Although he attributes his daughter's death to the woman's carelessness, blaming himself only for having left the baby in her care, he implicates himself through the unnecessary prescription of the cold bath, the indirect cause of her death. As a man of science, Withers is misguided in the decisions he has made, both those leading up to his baby's death and, here, his well-intentioned but counterproductive treatment of Zoriada.

The weeping Zoriada's response underscores the metatextual irony of Withers's story: she replies, ambiguously, "[A]n invisible power has brought me here, and all I can do shall be done" (I, 47). That invisible power is the novelist choreographing the interwoven destinies of her characters. Gibbes realizes her providential role in the relationship between Withers and Zoriada through one of the most unlikely of the novel's many plot twists: the Withers's lost son—having wandered off as a young child during the crisis with his sister—is discovered late in the novel to be Mims, the sea captain who rescued Zoriada and who turns out to be father of Edmund, the man Zoriada will marry by the novel's end. Gibbes's design is thus larger than the characters' or reader's ability to deduce; yet, paradoxically, Gibbes grants her heroine an active subjectivity. Zoriada's cryptic "all I can do shall be done" thus suggests her potential power even over the trajectory of the novel.

In her aptly titled chapter, "The Scrutiny," Gibbes's novel returns to Zoriada's room through a contrasting though equally failed "scrutiny" to that of Dr. Withers. Gibbes portrays the hegemony of western hermeneutics as both absurd and unethical through the malevolent Parson Swinborne, a leering voyeur. Enlisting the aid of Mrs. Leland to break into Zoriada's room, Swinborne arms himself with a magnifying glass "which hung pendent in the true coxcomb style, by a black ribband from his neck" (I, 112). Through his ludicrous hermeneutics, Swinborne vainly examines what he ridicules as Zoriada's strange "gimcracks" and "globuses"; the latter perhaps refers to the model Withers had identified as her "armillary sphere," Swinborne's "*globuses*" instead reducing the sacred relics of this globe-traveling but reserved stranger to a Latin term that hints at the misogynistically conceived condition, "*globus hystericus*," a phallocentric term for anxiety paralyzing one's voice (I, 109). Regarding what he perceives as Zoriada's lack of vanity, Swinborne assumes she is "either the most insensible, or the most artful of females"; Mrs. Leland, who is by contrast determined to prove Zoriada's innocence, articulates her hermeneutical difference from Swinborne in terms of the binary of female stereotypes available to them: "See . . . who is wrong and who is right, I for believing her an angel, or you for taking her for a—" (I, 93, 110). Her incomplete sentence suggests the term is taboo, for the choice is not angel/devil but angel/whore. Proving neither of these categories is accurate in knowing her heroine, Gibbes challenges the traditional gender duality, thus placing her at the forefront of the late eighteenth century's nascent women's rights movement.[12]

Swinborne's violation thus takes Doctor Withers's well-meaning if invasive scientific scrutiny of Zoriada's body-as-machine to a predatory level, the connection of Zoriada to the objects in her room underscoring the puppet-like objectification against which the novel works. Unlike Withers, who had not known what to make of the art in Zoriada's room except to assert that it was her creation, Swinborne is determined to discover Zoriada's identity by interpreting her drawings. One appears to him to represent the "black hole of Calcutta" including "a prison with the dead and dying, lying before it" which he later assumes proves her being party to the "scene of carnage" (I, 112, 115); another is "a house, with one side in flames, and on the other a slaughter of helpless individuals; blood, blood, Iago, a dismal scene" (I, 113).[13] His allusion to *Othello* says more about Swinborne's projection of faulty hermeneutics onto the drawing than it does about Zoriada: only late in the novel do we learn, contrary to Swinborne's assumption that the drawings verify her guilt, that they depict her having escaped when her family was murdered and their home burned. The latent revelation contrasts Swinborne's projection of his own capacity, like that of Iago, to bring about the downfall of an innocent, suggested by the culminating representation of a girl with an urn; Swinborne's descriptions of the symbols up to that point are of a devouring feminine principle: the "black hole" of Calcutta; the prison; the house in flames. That the reader is no cannier than Swinborne underscores both the providential role of Gibbes and her withholding of Zoriada's subjectivity until Volume III, in which the meaning of the girl with the urn becomes clear: Zoriada's desire to keep her family's remains with her puts her at odds with the burial practices of the village. By this point in the novel, the invasion of Zoriada's room suggests a double projection in which the objects that are studied to understand her come to inform her with a body viewed as a puppet-like shell.

Zoriada's identity nevertheless eludes the aggressive hermeneutics of characters and reader. In this pivotal scene, Swinborne's investigation of Zoriada's possessions is but prelude to his discovery of the most mysterious object in the room: a "small manuscript, in the Oriental language, intitled [*sic*] *Zoriada*" (I, 114). Finding it impenetrable, Swinborne plots to have the manuscript translated, convinced that it holds the key to Zoriada's enigmatic drawings that he already assumes, incorrectly, point to "her being a party in [its] scene of carnage" (I, 114, 115). Swinborne's objectification of Zoriada is thus a grotesque parody of the hegemonic, masculinist dualism at the heart of Enlightenment epistemology. That the title of the manuscript is the same as the title of the novel underscores Gibbes's metatextual irony, a reminder that the reader participates in the villagers' aggressively misguided "scrutiny" of her heroine. Yet Gibbes portrays the villagers through multiple levels of satire and irony, here balancing Swinborne's caricaturish swindling with the more nuanced flaws of Lord Drew, whom Swinborne has enlisted to steal the manuscript.

Unlike Swinborne, Drew has the potential to make ethical choices and thereby becomes a potential suitor to Zoriada. Gibbes creates in Drew a character who, not only wishing to earn privilege but, hoping his ancestors earned their entitlement, hearkens to the late eighteenth-century shift towards social mobility: "[I]t was the fame of well doing, and well-meriting, that he must conceive ennobled his forefathers" (I, 139). Nevertheless, "fashion" rather than "natural depravity" has turned Drew into a "libertine," for he "had not the firmness either to shun or detect vice" (I, 142). His failure in the hermeneutics of detecting Swinborne's vice, in particular, complicates his growing love for Zoriada. That Drew becomes concerned about reputation, referring to it as "the world's dread laugh" at the "man of such daring, as to marry a stranger in the land," takes on retrospective irony: Edmund Mims, forming the third point in this love triangle, was born in India, contrasting Drew who wishes "he had been born on the banks of the Ganges" so that he could ask "her hand with confidence," unhindered by social mores from which Swinborne ironically leads Drew astray (I, 147). Playing into Swinborne's plot, Drew has him go to London to seek out Mrs. Quinbrook, beginning a motif of romance as business that becomes increasingly complicated in Volumes II and III. Agreeing to gain information about Zoriada so that Drew can marry her, Swinborne baldly states, "I accept the terms of purchase" (I, 152).

Swinborne's aim in the business of manipulating the marriage between Drew and Zoriada is no less than "a grand source of knowledge," to bring about "the downfall of the loveliest girl on earth" (I, 184). His plot works against the providential plan of Gibbes and the emerging power of Zoriada's subjectivity: Swinborne "conjure[s]" Drew to write to him by every post, Gibbes ironically echoing Swinborne's own projection of evil onto Zoriada; Swinborne's plotting takes an increasingly dark turn through the ironic connection between his insistence on detecting Zoriada's witchcraft and the nadir of his actual evil (I, 152, 153). Plotting the kidnap and rape of Zoriada, Swinborne tells a fellow conspirator of his plan to abduct her to the home of a deaf man and "almost blind woman" who drink; thus, Swinborne tells him, "we may, without danger of detection, do whatever we please"; his choice of the hermeneutical word "detection" increases the ironic divide between the elusive Zoriada and the predatory masculinity of Swinborne's self-aggrandizement (I, 186). Asserting her subjectivity with increasing power, Zoriada ultimately becomes author of her own destiny. The manuscript, the novel's symbol of metanarrative hermeneutics, goes unresolved, however, still untranslated as the novel ends. The reader is left to speculate upon what is knowable through the limited paradigm of western epistemology.

III

Even as Gibbes leaves no ambiguity about her outrage over the abuses perpetrated by the empire of "Barbarous England," the novel's multiple points of view vis-à-vis colonial abuses further problematize the binary of empire and colony in Volume II. The entitled though morally educable Drew recognizes, for instance, that "civilized Great Britain does have cruelty she likens to destruction of her family in India" (II, 13ff., 17). Thus, though India is the locus of carnage that Zoriada has fled, Gibbes never specifies whether the perpetrators of the violence against Zoriada's family were indigenous Indians or colonists, the novel ultimately revealing the mixed bloodlines that deconstruct the empire/colony binary. Far from being the safe haven to which Captain Mims appears to have spirited Zoriada, the English village is filled with peril, as Swinborne's sequence of abuses represents with increasing aggression, from his symbolic violation of her room and projection of superstition onto the objects there to his attempted kidnap and rape of Zoriada.

As this aggression intensifies, however, Zoriada's voice emerges, a counter to the villagers' desire for knowledge through hermeneutical investigation as a means of hegemonic control. Gibbes superimposes references to Genesis onto their craving for knowledge, simultaneously representing through Zoriada's relationship to knowledge a counternarrative to the Judeo-Christian story of the desire for forbidden knowledge as original sin. Gibbes makes the connection among

Enlightenment hermeneutics, gender, and colonialism yet more pointed when Drew, leaving the village for "the great city," invokes *Paradise Lost* (11); ironically Gibbes gives Drew Eve's words when, departing "with the same lingering pace our first parents are described to have quitted the primitive spot of human existence . . . , [he] seem'd to exclaim[:] 'And must I leave thee, paradise / For ever leave thee!'" (II, 36–37). Eve speaks the first line in this passage from *Paradise Lost* when Michael tells the fallen couple that they must leave Eden; the irony here is that Drew must depart the paradise that the village has become because of Zoriada's presence: she has imbued it with innocence that transcends the biblical binary of pre- and postlapsarian Eden. That Swinborne's *modus operandi* is "insinuating" lexically connects him to the serpent in Eden, thereby echoing and deepening the connection between the desire for knowledge of Zoriada by the villagers, for whom it is axiomatic that "Curiosity was nevertheless the sin in paradise" (I, 84); the villagers are thus implicated in these Miltonic terms, their agrarian paradise deconstructed as a postlapsarian world where knowledge is a means to appropriation through intrigue, spying, and coercion, paradoxically distinguished from the nondual innocence of Zoriada.

Returning to the business motif introduced earlier through Drew and Swinborne's plotting, Gibbes shows that even Doctor Withers is not exempt from the taint of Swinborne's manipulation; he recommends that Drew "make an immediate journey to London" as a means to "promote your tender interests," namely, "the true claims of this lovely stranger that has stole all our hearts" (II, 33). Gibbes's irony here emerges through the word "stole": taking the tired Petrarchan trope of the male subject having his heart stolen by the female object, Gibbes reverses the gendered association, Swinborne having literally stolen Zoriada's manuscript (II, 42).

Just as the plotting to win Zoriada's hand in marriage intensifies in Volume II, so too does the attempt to coopt Zoriada's story. In his desire to make her attractive to Drew, Captain Mims tells Drew Zoriada's story before Zoriada herself does. Ironically, however, Mims's story is a graphic, first-hand account describing the aftermath of the massacre of Zoriada's family and destruction of her home in India. In Mims's account, a wounded man approaches him for help, telling Mims he was servant to a family that has just been butchered, though "whether from motives of malice or plunder cannot be ascertained"; he tells Mims of "treasures . . . buried deep in the earth . . . with an Arabic inscription" to guide him (II, 47). The man says that the "eldest child of the family must have escaped," urging Mims to "seek her, preserve her" so that he can die in peace (II, 48). In graphic detail Mims then tells Drew of the murder scene he discovers: "Tracks of blood were our clue to the house, where seventeen persons lay butchered, and though dead, were not cold, and even defaced as they were, and their cloaths [*sic*] drenched in gore, we could distinguish that two of them were ladies, a young and an old gentleman, three children, and the rest attendants" (II, 49).

By contrast to Mims's detailed description of the brutal scene, Zoriada, as the only one of its victims besides the servant to escape, cannot describe it from such a graphically objectified perspective; instead, she nostalgically describes her lost home and family as "this garden" that had "smiled" (II, 53). Gibbes has Zoriada transform the garden imagery from the earlier perspective of the villagers' quest for knowledge as original sin. Here, Zoriada instead portrays the garden of her Indian home as communal in its state of innocence: "affection and friendship glowed in every heart; the conversation, how delightful!" (II, 53). The innocence of Zoriada's garden thus transcends the prelapsarian/postlapsarian binary of Genesis not only because it was communal but, as we come to learn, racially and culturally mixed.

The fate of the urn containing the remains of Zoriada's family parallels the villagers' attempts to appropriate Zoriada's subjectivity in Volume II. Mims promises Zoriada to give her family a "funeral pile erected for them after the Eastern manner," after which he would "collect their ashes into an urn, to be deposited in some suitable sepulchre in England" (II, 54). Mims, however, confides to Drew that "this idea of burning the house, [*sic*] and its unfortunate inhabitants suggested itself to me as the most effectual means to detach her mind from India; where, if she continued under my protection, she could not long reside, and it had in a great degree the desired effect" (II, 55). Mims's assumption that he can achieve catharsis for Zoriada through immolation of her Indian past is laden with irony, especially as he tells the story following the villagers' earlier impressions of Zoriada as an embodiment of pathos.

Following Mims's story, Drew marvels at the "mystery and *inexplicability*" of Zoriada, Gibbes's italicized term stressing both the inability of the captain to tell the story and the suitor to apprehend her story from the elder man's perspective (II, 60). The immolation of the dead and burial of the remains become literally and metaphorically complicated when Volume II returns to Zoriada's initial request to Mrs. Quinbrook to protect not only her but her "whole family . . . contained in this precious urn, which shall be deposited in my own chamber" (II, 67). In spite of Mims having promised Zoriada to respect Indian burial rites, he chastises her for ignoring his "judgment and decision: that I have held that urn sacred, I need only refer you to my conduct for a testimony, but when I proposed bringing it over, I little imagined you would have violated, you must allow me to call it the rights of the dead, or robbed the sepulchre of its proper deposit" (II, 68). Only now does the reader learn Zoriada's wish to keep the urn with her family's remains in the room that Swinborne desecrated. That Zoriada has included the urn in her chamber, her own created microcosm, until she is ready to bury the urn, contrasts Mims who "decides" it should be in a sepulchre; "taking advantage of her silence," Mims cruelly "conveyed the melancholy memento out of her sight" (II, 69).

The literal burial takes on metaphorical significance when, contrary to the Captain's own desire for Drew to win Zoriada, Mrs. Quinbrook, rooting for Edmund, Mims's son, decides to "bury . . . deep in her own bosom" her "design . . . until an opportunity should mature them, and prepare a way for their happy prosecution" (II, 72–73). That design, to "counterplot" the business-transaction deal between Mims and Drew that, she is convinced, will "manoeuvre the young lady out of her heart," is laden with multiple layers of irony: the "opportunity" to mature her designs has metanarrative implications, reminding the reader that Gibbes is behind Quinbrook, maturing the plot and wresting control of it from the machinations of Swinborne, Drew, and the Captain. The language of burial is thus ironic in the double use of the term immediately preceding this moment: the exhuming of Zoriada's treasures and the burial of her family's remains. By having Mrs. Quinbrook keep her motives hidden or "buried" thus suggests the doubleness of the novel's conflicting burials: of treasure—here, knowledge—to be exhumed for the right time but potentially repressed or threatened. A third connotation of burial is added in this phase of the imagery's development: the agricultural suggestion of planting the seed until it is ripe. Yet this connotation is compromised linguistically as well, since the designs will have "a way for their happy prosecution," an ironic term for fruition as it suggests the litigation of a crime (II, 73). Drawn into the hermeneutics of the novel and wishing for the happy ending from Mrs. Quinbrook's point of view, the reader is thus implicated in the crime against Zoriada who, behind all this espionage, silently waits for her own opportunity to assert her subjectivity. She knows nothing of either party's machinations as they plan competing "nuptial festival[s]" (II, 76).

Gibbes creates a counterbalance between burial and the exhuming of knowledge when the novel hints at the truth of the biological relationships among the villagers and Zoriada. Here, Captain Mims identifies as his parents the couple who raised him after he ran away from his biological parents, the Witherses, following the death of his sister; Mims tells Zoriada and the Witherses that he had to "bury" both parents—Anglo-Indians—in Calcutta, where he was raised. Mrs. Withers senses her "lost child's accents" in Captain Mims's story; simultaneously, Crosby beholds the "living image of [his] lamented brother" in Zoriada, these plot twists returning in Volume III when it is revealed that Mims was the lost son of the Witherses, adopted by Anglo-Indians, and Zoriada is the daughter of Crosby's Anglo-Indian brother (II, 82, 83).

Mrs. Withers, not yet knowing Captain Mims is her son, is convinced that Edmund Mims, her grandson, is her lost son; not realizing she is merely a generation off in her hermeneutical reasoning, Mrs. Withers says, "His age does not agree with the period of our misfortunes; and yet his features, his features, are our child's!" (II, 119). Her inability to verify her observation leads to "dreams that tortured her; she insisted upon it there was some concealment about the young fellow," forcing Edmund to reveal his identity to Zoriada (II, 120). Gibbes frames the revelation of Edmund as the romantic hero by giving him a Romantic sensibility as he takes in the mysterious beauty of the monastery: "Mims was no stranger to all its intricacies, the rock, the grotto, the cave" (II, 137).[14] Through the cave's "subterraneous passage" he leads Zoriada, bringing her "into full light at the foot of the declivity, where a prospect of the seas was enchanting, and the cave of his former residence and asylum, beyond description, antique and striking" (II, 137–138). Zoriada assures him, "I am above the little customs of my sex" when he asks if she hates him (II, 139). Her response to the revelation underscores her subjectivity and his acquiescence: "Rise sir," she says, adding, "Accept my hand to assist you to rise" (II, 140). She says to herself, "He deserves my hand," observing her heart "had decided in his favour, before I knew his claim to it" (II, 140). Beyond the complicated selfhood that the pseudonymous Zoriada represents, even to herself, here her heart represents an intuitive knowledge that transcends the ineffectual hermeneutics of the villagers who nevertheless appear capable of embracing the intuitive, as when Mrs. Withers knows that Edmund is her relation even though the complexities of the events don't yet make sense.

Volume II closes the framing imagery of Genesis, now from the point of view of Drew, who has just discovered Mims is courting Zoriada: with "deceitful calm," Drew returns to "behold Zoriada in the garden, chatting, with apparent delight with his rival"; he continues, "The Devil, when he beheld the first pair in Paradise, had not more envious or more torturing sensations" (II, 145). As the model of a new Romantic masculinity, Edmund contrasts Drew, having earned his worthiness in the eyes of the woman who chooses him not only because his "understanding [is] cultivated to [her] taste"; even his lower social class makes him more desirable, as Zoriada tells Drew: though Mims is his "inferior" in "the world's estimate," his "spirit is gentle, his manners engaging, his understanding cultivated to my taste, and need I speak the consequence, thus, my Lord, you find me a woman of my word; and it now remains for you to prove yourself a man of honor" (II, 148). Equally interesting in the context of Edmund's attributes is that Zoriada identifies with him as "an East Indian by birth," regardless of his English ancestry (II, 146). Gibbes nevertheless complicates Drew as a foil to Mims by giving a multiplicity of conflicting perspectives on him: while Zoriada does not "read" Drew's behavior towards her as dissimulation, thinking he is "subscribing to her wishes," the Witherses "could read in the sullen brow of Lord Drew, sentiments inimical to general tranquility" (II, 150). Mrs. Quinbrook, in spite of wanting Edmund to win Zoriada, sees Drew as "very distant and reserved" (II, 151). This contrast between the conflicting views of Drew and Mims, idealized precisely because he is identified with Zoriada as sympathetic and East Indian, culminates when Drew challenges Edmund to a duel. Edmund is "petrified," a reaction that highlights the new Romantic masculinity, not virile but sentimental (II, 154).

Gibbes represents these two men fighting over Zoriada, Drew objectifying Zoriada as "the beautiful East-Indian," as ludicrous since the choice will ultimately be Zoriada's (II, 154). Zoriada, by contrast, uses language that points to the hypocrisy of British judgement of indigenous East Indians as barbaric when she reasons to herself that Drew seeks a "human sacrifice" as the duel's desired end (II, 155). Volume II thus ends with Zoriada successfully interceding on behalf of Edmund, whose identity she reveals as the son of her rescuer, Captain Mims. In spite of having been corrupted by fashion to the point of engaging in this barbarity, however, Drew has a "natural goodness of . . . heart" that had "repeatedly suggested to him the violence, and injustice of his proceedings" so that Zoriada is able to reason with him (II, 156).[15] In the context of the violent underpinnings of the novel, Gibbes's darker condemnation of the abuses of the powerful can be discerned underlying Mrs. Leland's comical remark on the difference between a gentleman and a low-born man: for the gentleman, "nothing but cutting one another's throats can content them" as opposed to "the good people of low degree" who "can box away their anger (III, 3).

IV

Volume III builds towards a paradoxical conclusion by intensifying the conflict between the reader's desire for a happy resolution to the marriage plot and the multiple perspectives of the characters that undermine such a prospect.[16] Thus, while Zoriada expresses her pity for Drew regarding his rashness, Gibbes uses Mrs. Leland to refer to the artificiality of a tidy ending through metafiction: Mrs. Leland remarks that "no stage play can out do Madam Zoriada's behavior, and yet my mind had its mis-givings, for a second time, for as certain, as I once believed, over-learning had killed her, so certain, did I this day think it had driven her beside herself" (III, 8). Referring to the duel as a "transaction," thereby picking up the thread of the business motif in describing marriage, Withers condemns Drew: "I hate an assassin . . . however dignified" (III, 9). Yet Withers blames women for duels, saying that if "no one woman" were found "capable of smiling upon the successful murderer" it would put an end of duels (III, 10).

Gibbes balances Withers's masculinist comment through the servant, Martha, who rages against the "barbarous" Lord Drew, underscoring the novel's indictment of class inequities and the hypocrisy of the aristocracy (III, 34). Her diatribe undercuts any sentimentality about Zoriada as heroine, since Martha's main concern is losing her "place." With unabashed honesty she lets Drew know the source of her anger is the worry about losing her livelihood, coming even before her concern for Zoriada: "now you have killed her, and turned me out of my place, I must not tell you of it, because you are a lord; but I say, noble is, that noble does; and if a prince, nay a king, was to act the part of a villain, I would twit him with it to his face, though he hanged me for it" (III, 35). Through Martha, Gibbes reverses the gender binary of rational men and seductive women, Martha reducing Drew to a "conjuror" who "bewitched us all," his castle, she imagines, "full of unhappy women" (III, 36). Crosby, who walks in on her upbraiding Drew, reveals his own liminality between the two extremes they represent when he notes that Martha is "one of the best but most uncultivated of hearts" (III, 39).

The novel complicates Zoriada's assertion of her subjectivity yet more strongly with this emerging focus on the relationship between gender and social class. It is in this context of class that Zoriada proposes to Crosby that he marry her to Edmund and that they elope: "I shall be one and twenty next week, Edmund is already three and twenty. . . . The superiority of my fortune, it seems, creates a delicacy between us; could Mims reverse our conditions, I know he would not cease soliciting me until I bestowed my hand upon him" (III, 64–65). Yet it is not merely a matter of the reversal of their stations that is at stake in her hints about the nature of the "fruits brought with [her] from India" (III, 67). When Crosby urges her to tell him her story, she says he must wait until Edmund is there, for "the story is too melancholy to be told more than once" (III, 67). In a novel whose details of dead and missing children and parents accrue with each character's story, the metatextual statement is ironic, putting the reader in a double bind regarding his/her suspense at wanting to know the truth of Zoriada yet finding repeatedly a heart-numbing regularity with which tragedy unfolds at every level.

Zoriada's voice reaches its full strength in her letter to Edmund Mims proposing their elopement. She asserts her authority by appropriating the language of business: "No transaction was ever more privately, or silently effected" (III, 71). Meanwhile, Captain Mims discovers that Mrs. Quinbrook has betrayed him by allowing Mims to see Zoriada and plots to send Edmund to India by tricking him to go aboard a ship to have dinner with a "brother captain" (III, 82). That Gibbes not only sends her Indian heroine to England but has her romantic hero, an English national, sent to India to be trepanned, or kidnapped as a slave, deconstructs yet another binary along with those of culture, social class, and colonialism: active male/passive female. While Edmund is thus trepanned, Captain Mims reveals that he is the Withers's son after news arrives explaining how he came under the care of his adoptive parents, Stephen and Jane James, who had found the boy thirty-eight years before. When Mims then announces that he has just shipped the Withers's grandson, Edmund, off to India, Drew decides to rescue his rival, underscoring that he is essentially good if narcissistic, corrupted by education: "it will be an act worthy of humanity," diminishing the idealism of the statement by claiming, "and give lustre to my character" (III, 95).[17]

Zoriada's final actions in the novel emerge as authoritatively as do her words when she gives her inheritance to the farm dwellers "who shall hold it rent-free." She claims, "No one has the power . . . I am of age, I am a wife; but my husband's heart is in unison with my own, and he will promote every good work, I am inclined to engage in" (III, 128). Though the novel appears to move towards a happily-ever-after, it is tinged with the "solemn ceremony" in which the urn containing the ashes of Zoriada's family, "a whole household," is interred. The inscription eulogizing the family "who fell victims to the barbarous rapine of a set of ruffians on the banks of the Ganges" reduces this "unfortunate family" to one half of a binary—"English extraction"—implying the other half, the "barbarous rapine," is indigenous (III, 138). Yet the novel itself deconstructs such a claim to "English extraction" with Zoriada's story of her mixed bloodlines in the final chapter. When Lord Drew produces the manuscript stolen by Swinborne, Zoriada responds, "When I have a little overcome . . . these tumults, I will endeavor to put the story into an English dress, for the general entertainment" (III, 149). That such a time for her to translate her manuscript into English would be premature at the end of the novel suggests that the translation would run counter to the English hegemony of the urn's inscription.

Gibbes saves for the novel's final chapter Zoriada's narration of her story in her own voice. Zoriada describes herself as "half blood," a category underscoring her paradoxical position as both subject and object; Zoriada's is thus a selfhood at the nexus of the waning binary system extolled during the Enlightenment and the nascent exploration of nondual epistemology associated with Romanticism (III, 159). Zoriada reveals that her father was an Anglo-Indian gentleman, employed by the East India Company; imprisoned due to the earlier transgression of his brother, Crosby—now, therefore, identified as her uncle—her father was freed from prison by a Muslim whose daughter he would marry, producing Zoriada. Crosby, now revealed to be Zoriada's uncle, realizes that she is "the offspring of a brother I have undone" (III, 160–161). The culmination of the novel's deconstruction of the western binary system is thus its imagining an India that is half-blood: Zoriada embodies a feminized India sharing her riches both east and west with the downtrodden of England, teaching a western patriarchy about the importance of its own subjectivity, and redeeming those who have been corrupted rather than being plundered by mercenary imperialism, thus entangling the strands of ancestry among caste, race, and religion. Gibbes leaves the manuscript untranslated and therefore the greatest mystery of the novel remains unresolved, suggesting that the unknowability of the east is a deeper strain than the Enlightenment belief in human potential to know the secrets of the universe.

Gibbes paradoxically asserts her own authority by subverting it as the novel moves towards its ultimate refusal to define selfhood for her protagonist, whose real name is never revealed nor can the manuscript reveal her relationship to the binary system upon which the English language of the late Enlightenment that would translate it operates. Leaving the Arabic untranslated is a triumph over the villagers' and reader's attempts to classify her, as an English translation would, according to the binaries that have driven the novel's plot: good or evil, rich or poor, active or passive, colonizer or colonized, or, implicitly, virgin or whore. Standing as a proto-feminist critique of orientalism, Zoriada the heroine has overcome western hegemony manifested as the denial of her own subjectivity; so too has *Zoriada,* the manuscript as textual symbol of eastern epistemology, overcome the villagers' attempted cultural appropriation. By positioning Zoriada in relationship to the provincial villagers, Gibbes not only signals the late eighteenth-century ambivalence underlying the revolt against Enlightenment thought; the novel gestures ahead to modern and even postmodern representations of unknowability by making a distinction between the fashioning of selfhood through one's story and selfhood that resists the laws of time and space upon which narratives are constructed. In spite of Gibbes's providential design and adherence to the sentimental formula of the marriage plot, she gives Zoriada the authority to choose the man she will marry; even more indicative of the autonomy Gibbes grants Zoriada over the quest for knowledge, by allowing her to keep the manuscript untranslated to the end and her name undisclosed, Gibbes grants her heroine authority beyond Gibbes's own authorship.

NOTES

Chapter title: Phebe Gibbes, *Zoriada, or Village Annals* (London: T. Axtell, Royal Exchange, 1786), 177; *Eighteenth Century Collection Online.* All references to this source are in the text.

Retrieved from the GALE group website at https://www.gale.com/c/eighteenth-century-collections-online-part-i (last accessed October 27, 2019).

1. For a recent discussion of the oriental tale genre, see Ros Ballaster, "Fakiry: The Oriental Tale" at http://writersinspire.org/content/fakiry-oriental-tale, licensed as Creative Commons BY-NC-SA (2.0 UK).
2. Though *Zoriada* was previously attributed to Anne Hughes, Gibbes, who wrote twenty-two books between 1764 and 1790, claimed the novel in 1786; her authorship is supported by her characteristic style seen in elements such as ironic juxtapositions; ubiquitous allusions to the English poetic tradition, with references from Shakespeare, particularly *Othello*, Milton's *Paradise Lost* and Thomas Gray; and its metafictional sophistication. *Zoriada* includes a nod to Sophia, the protagonist of *Hartly House Calcutta* who, like the secondary character of the same name in *Zoriada*, writes letters to her friend Arabella though, unlike her namesake in *Hartly House*, Sophia here becomes a tool for the narrator to idealize the feminized East (III, 98). See note 3, below, for more on citations from this text.
3. The manuscript uses the spelling *Zoriada* rather than the traditional *Zoraida*. Whether or not the transposition of vowels is an idiosyncratic choice or an error, the essay follows this and other irregular spellings in the manuscript, bracketed for clarification when necessary.
4. A case in point is Coleridge's development from his early poetics founded on Associationism, the final phase of Enlightenment philosophy, to his outright rejection of it as he embraced German transcendental philosophy. For a recent discussion of his conflict with Enlightenment materialist philosophy, see Richard Berkeley, *Coleridge and the Crisis of Reason* (New York: Palgrave Macmillan, 2007).
5. For a detailed discussion of the sentimental genre, see Janet Todd, *Sensibility: An Introduction* (New York: Methuen, 1986). I discuss Gibbes's challenge to the genre in *Hartly House* through the "orientalizing" of her English protagonist, Sophia Goldborne in my book, *British Women Writers and the Asiatic Society of Bengal, 1785–1835: Re-Orienting Anglo-India* (New York: Ashgate, 2014).
6. See Christina Lee, "The Legend of the Christian Arab Madonna in Cervantes' Don Quijote," *Revista Canadiense de Estudios Hispánicos* 32, no. 1 (Autumn 2007): 105–121, accessed March 31, 2019, https://www.jstor.org/stable/27764178?seq=2#page_scan_tab_contents.
7. Mary Tighe's early death, for example, was attributed to her desire for literary fame; in a letter, Felicia Hemans quoted Tighe's mother saying, "Oh! my Mary, my Mary! The pride of literature has destroyed you!" (quoted in Norma Clarke, *Ambitious Heights: Writing, Friendship, Love: the Jewsbury Sisters, Felicia Hemans, and Jane Welsh Carlyle* [London: Routledge, 1990], 50–51).
8. *Encyclopaedia Britannica*, s.v. "Lute," Encyclopaedia Britannica Inc., 2019, accessed March 31, 2019, https://www.britannica.com/art/lute.
9. Ancient Origins, "Armillary Spheres: Following Celestial Objects in the Ancient World," accessed March 31, 2019, http://www.ancient-origins.net/artifacts-ancient-technology/armillary-spheres-following-celestial-objects-ancient-world-004025. For a fascinating lesson using an actual armillary sphere, see Chris Parkin's demonstration for the Museum of the History of Science, accessed March 31, 2019, https://www.youtube.com/watch?v=AaWuJHQL-bQ.
10. Adrian Johnstone, "Galileo and the pendulum clock," accessed October 6, 2019, http://www.cs.rhul.ac.uk/~adrian/timekeeping/galileo/.
11. https://plato.stanford.edu/entries/scientific-method/ (accessed March 31, 2019).
12. The novel anticipates Mary Wollstonecraft's 1792 *Vindication of the Rights of Woman*, considered the first detailed analysis of the masculinist system of women's education in England.

13. The allusion to Iago may be further evidence of the connection of this novel to *Hartly House*, in which references to *Othello* are prominent, thereby strengthening the identification of Gibbes as author of both novels.
14. The term "romantic" is overdetermined, with a long history and a variety of connotations including "erotic" and "transcendent." I use the capitalized Romantic to indicate the ideology of the revolutionary period in England; the subject of Romantic masculinity, discussed below, involves the shift from earlier definitions of manliness connected to martial virility to the traditional feminine quality of emotion that came to characterize Romantic poetry. For his pioneering discussion of shifting definitions of masculinity during the period, see Marlon Ross, "Romantic Quest and Conquest," in *Romanticism and Feminism*, ed. Anne Mellor (Bloomington: Indiana University Press, 1988).
15. Indeed, his name comes to suggest that though he "drew" his sword, he has the wherewithal to withdraw it.
16. This volume appears unedited in comparison with the markups and corrections in Volumes I and II. Because the chapters appear mismarked at a certain point, I give both the printed chapter number and what is probably intended based on the sequence from the beginning of the volume.
17. Gibbes represents the cruelty of Captain Mims towards Edmund, whom he is sending away at the peril of Edmund's new wife, through an ironic literary reference: Captain Mims quotes Colley Cibber's 1699 revision of Shakespeare's *Richard III*: "They will bend, indeed, indeed, but he must strain, that cracks them" (III, 115–116). In *Richard III*, Gloucester's lines are an aside when he appears before Lady Anne. This technique is ubiquitous in *Hartly House*, as noted above.

Part Three

ORGANISMS

8

SCHELLING'S UNCANNY ORGANISM

Stefani Engelstein

IN HIS 1919 ESSAY ON the uncanny, Sigmund Freud cites pages of definitions of the word and its cognates from foreign language dictionaries as well as the nineteenth-century German dictionaries of Daniel Sanders and the Grimms. From these pages he culls two points. First, he notes the reversal of meaning by which *heimlich*, or homey, comes to coincide with its negation, *unheimlich*, uncanny, in a process by which the domesticated transforms itself into the sequestered, secret, and buried. Second, he isolates a particular definition offered by Friedrich Wilhelm Joseph Schelling, who calls the *unheimlich* "the name for everything that ought to have remained . . . secret and hidden but has come to light."[1] Indeed, Freud will go on to adapt Schelling's definition into his own, locating the source of the uncanny in the return of the repressed, and, in fact, in any reminder of the mechanisms of repression and the existence of the dark recesses of the unconscious within that most intimate domicile, the psyche. The reference to Schelling may seem merely incidental, but a return to Schelling on his own terms reveals an involuted view of the workings of nature, the world, and the human intellect that renders all internally organized, self-regulating systems distinctly uncanny, paving the way, not coincidentally, for a Freudian unconscious, but also perhaps for a less human-centered ontology, one that has made Schelling an increasingly popular touchstone in the twenty-first century.

Freud's Schelling quotation was taken from a set of lectures entitled *Philosophy of Mythology*, delivered in 1842 and not published until after Schelling's death. During the last fifteen years of his life, Schelling immersed himself in investigations of religion, revelation, and mythology. He thus returned to his earliest preoccupations. The brief document, "The Oldest Systematic Program of German Idealism," thought to be jointly authored by Schelling, Hegel, and Hölderlin in 1797, appealed for a "new mythology . . . [which] must be in the service of ideas, it must become a mythology of *reason*,"[2] and Schelling's 1800 *System of Transcendental Idealism* repeats the call.[3] At this early point in his life, Schelling presents mythology as a medium through which philosophy in the ancient world was intertwined with poetry and through which it could be again, thus re-uniting the subjective and objective in ways to which we will return below. In his late-life lecture, Schelling's definition of the uncanny addresses the relationship of Greek mythology to "oriental" mythologies and to mysticism:

> Greece has a Homer precisely for this reason, because it has mysteries, that is, because it succeeded in fully vanquishing and repositioning into the interior, i.e. into secrecy, into mystery (from which it had originally emerged) that principle of the past that was still on the surface and dominant in the oriental systems. The clear heaven that wafts over the Homeric poems could unfurl itself over Greece only once the dark and bedimming force/violence [*Gewalt*] of that uncanny principle (uncanny is a name for everything that ought to have remained secret, hidden, and latent and emerged)—that ether which vaults over Homer's world could only unfurl itself after the force/violence of the uncanny principle that dominated the earlier religions was precipitated into mystery . . . after the actually religious principle was hidden in the interior and left the spirit fully free on the surface.[4]

The orientalizing discourse in which Schelling participates here is central to the story of Europe's definition of itself as purportedly modern through a projected temporal differentiation mapped onto a geographical one.[5] This constructed divide is projected onto a contrast of forces, one dark and one light, one internalized and the other externalized in a reciprocal and interdependent process that, significantly, mimics Schelling's metaphysics.[6] Such reiteration is in fact a necessary part of Schelling's system, which strives to open the world to human knowledge by including both subject and object within the same domain of interactive forces. Indeed, what first appears to be an analogy runs deeper. Schelling's two-pronged argument here vehemently rejects interpretations of Greek mythology as essentially mystical, on the one hand, and as allegories for natural phenomena, on the other. Unlike natural objects, the Greek gods enjoy unconditioned freedom of movement; they are neither heavenly bodies nor animals, but, as demonstrated by their human form, depicted in statuary in the act of striding (*AWS* VI:668), are free spiritual beings (*AWS* VI:664–665); in their various personalities and in their narratives of dynastic and generational transmission, they demonstrate individuation (*AWS* VI:662), so that they are "not only moral, but also at the same time historical beings" (*AWS* VI:664). The progression of religion here towards a stratification of free spirit from mystical depth mimics that of the progressive (*AWS* VI:665) development of consciousness, but also that of nature itself, in Schelling's nature-philosophical and idealist writings. Schelling's depiction of the progression of nature also moves towards more voluntary motion (*HKA* VI:198), towards increasing individuation (*HKA* VI:222), and finally towards history (*System*, 198–203).[7] Indeed, as Schelling explicitly states, near the end of the lecture series, "spiritual products [geistigen Erzeugnissen] . . . in their inner necessity and law-governed development, are equatable [gleichzustellen] with natural products [Naturerzeugnissen]" (*AWS* VI:683).[8]

If then, the development of spiritual products like mythological systems is not merely analogous to the development of natural products from planets to people but actually the result of the same process, how might we understand the dark first principle buried in the latter whose untoward emergence into knowledge would elicit a shudder? What is nature's uncanny for Schelling? And what is its place within humans, or of humans within it?

ORGANISM

The configuration of the relationship between humans and nature that emerged out of the Enlightenment entailed a paradox. On the one hand, the translation of natural phenomena into the regular and hence predictable patterns of natural law accorded the experimenter power over nature and promised great benefits through mastery. On the other hand, the generalizability of such laws subordinated humans, including the naturalist, to nature. Increased practical power correlated exactly with a loss of autonomy on a theoretical level, and stranded the naturalist in a mise en abyme from which an escape was hard to envision. The rampant debates over materialism and vitalism in the eighteenth and nineteenth century attest to the perturbation caused by this double-bind, even, or especially, if the sides of such a debate are nearly impossible to cleanly delineate.[9] While Caspar Friedrich Wolff's *vis essentialis* or Georg Ernst Stahl's *anima* invited charges of vitalist metaphysical sleight of hand from some, Julien Offray de La Mettrie's human machines served others as materialist objects of ridicule, even if each view was generally distorted in the process.

At the end of the eighteenth century, a shift occurred in the discourse on living beings in an attempt to reconcile the terms of this debate. Living things were reimagined as active, internally integrated systems uniquely organized as both law-governed and spontaneous, material without supervenient spirit and yet nonetheless able to initiate actions. This new configuration entailed a thorough redefinition of both the discourse and the terms of the debate through the establishment of a series of analogies between organism and world, on the one hand, and organism and artwork, on the other.[10]

The new constellation at the foundation of the modern field of biology has been most frequently explored through the concept of teleology, which emerged as a constitutive element of the discourse on living beings within the intense research and theorizing about generation or reproduction in the eighteenth century. The most prevalent theory of reproduction in the early modern period, called preformation, had held that each generation was encapsulated in the reproductive organs of the previous generation, one inside the other, all the way back to Eve, so that all living beings from the beginning to the end of history were produced in a single act of divine creation.[11] The competing theory of epigenesis, postulated by Caspar Friedrich Wolff and eventually formulated into the popular theory of a formative drive by Johann Friedrich Blumenbach, granted each organism a point of origin at the moment of conception. Explaining embryogenesis required understanding how an embryo progresses towards a specific bodily form—both the general form of the species and the individuated form that registers resemblances to progenitors. Such development appeared to be propelled, in other words, not by responses to a previous impulse but by a future goal, a telos. Moreover, as Immanuel Kant elucidated in the *Critique of Judgment*, progress in the life sciences rested upon the nonprovable assumption that each organ served a pre-identified purpose within the bodily economy as if it had been assembled for such a purpose. This future-oriented element of life differentiated it from the kinds of laws attributable to nonliving matter. Timothy Lenoir has called the resulting view of life vital materialism, while Robert Mitchell writes of experimental vitalism.[12] Teleology did not stand in isolation as the only distinctive feature of living beings theorized in the period, however. Moreover, the era arrived at its own term for the unique properties of living matter, not purely mechanistic and yet encumbered by the laws of matter, namely, *organism*. The word *organism* came into usage at the end of the eighteenth century not as a synonym for *living thing* but as a descriptive term for the particular teleological dynamic shared by integrated, reciprocally interdependent systems of parts.[13] An *organism* is thus a kind of mechanism that, however, also exceeds mechanism.

Schelling, in an allusive return to Plato in his 1798 *On the World-Soul: A Hypothesis of Higher Physics as an Explanation of the Universal Organism*, brought the still rare word *Organismus* into greater prominence in the process of developing a *Naturphilosophie*, a philosophy that acknowledged nature and natural law as ontologically foundational for human intellect, while still positing the possibility of spontaneity or will.[14] Like Fichte before and Hegel after him, Schelling here attempts to solve the problem of Kant's reorientation of metaphysics through epistemology. Kant provided a foundation for transcendental agreement on the phenomena of the world by positing consistent categories of apperception across all intelligent beings. This move provided an anchor for knowledge, but delegitimized metaphysics as beyond all knowledge and unmoored knowledge itself from its purported objects. Fichte attempted to reverse this move, reconciling nature and intellect by reconstituting the world within the unconscious positing of the absolute I, so that the activity of production replaces representation. This move weakens dualism, however, only to the extent that it absorbs the object into the subject. While Fichte thus locates a split between subject and object within the subject, Schelling instead attempts to unify subject and object by positing a fundamental duality, in fact a duplicity, within a single originary unity.

Schelling's system depends, like Fichte's, on activity, but this activity is not a human privilege alone. Intelligence, although unconscious, belongs to the system of nature from the beginning, and nature, far from consisting of merely dead matter, instantiates a fundamental interplay of forces. The inorganic, the organic, and the human thus belong to a single organization. The primary organization of nature emerges from the interplay of two forces, a positive and infinite force that strives to expand, and a negative force that hinders the first. The forces become perceptible only in their joint products, from their most elastic manifestation in light at one extreme (*Weltseele*, *HKA* VI:78–79) to the hard bodies that, while "the darkest of things," as Schelling names them in an 1806 essay that accompanied later editions of the "World-Soul" ("On the Relationship of the Real to the Ideal in Nature," *AWS* III:590), are still an instantiation of the foundational duality, that of infinite striving under the constraint of finitude. The repulsive force of expansion most readily perceptible in light is thus opposed to the attractive force of gravity most readily perceptible in massy bodies (*HKA* VI:80–81). Any given form of matter has reached a point of equilibrium or rest appropriate to the interplay of the dual forces that is expressed through it. This state of rest is what enables matter to persist, but is only relative (*HKA* VI:78–80); absolute rest could only result from the absolute dominance of one or the other principle—absolute cessation through the victory of the hindering force dominant in matter, or entropy and absolute dissolution through the victory of the primary expansive force. Within nature, in other words, lie two equally destructive tendencies, each held in check by an opposition that produces active and organized matter as a side-effect, a detour. Mechanism, or the sequence of cause and effect, describes a simple, forward-striving progression; *organism* is also a succession of cause and effect, which, however, "*enclosed within certain limits*, flows back on itself."[15] While mechanism is thus inherent to organism, organism is equally necessary for the duration of mechanism (*HKA* VI:69).

This organism is not confined to living beings, which, however, represent a particularly complex order of organism. The world itself as a whole also possesses an organism. The unification of organic and inorganic nature under a single principle and a single set of forces makes life available to the investigations of naturalists, that is, to knowledge and to explanation, in the same way that simple matter had been beginning with Newton. There are, however, also differences between the inorganic and the organic; both are built upon the same duality, but the living raises its oppositional reciprocal action to another level through ever more refined dualities. Already in the work's title, Schelling calls this principle of opposition within unity alternately *organism* and *world-soul*; it is self-directed, organized activity. The term *world-soul* is borrowed, as Schelling notes several times, from the ancients, primarily Plato's *Timaeus* (and secondarily his *Philebus*). Schelling wrote an elaborate unpublished commentary on the *Timaeus* in 1794.[16] The *Timaeus* revolves predominantly around the telling of an origin myth, but not one that belongs to the common mythological cycle populating Greek epic, lyric, and drama, although it does share several features of this mythology. The *Timaeus* features divinities, arising successively in history and acting freely. These divinities, however, inhabit a threshold between the most advanced plane of mythology described by Schelling in his lectures on mythology, on the one hand, and philosophy, on the other. Their personalities have moved from the individuated to the relatively abstract, and the acts of creation in which they participate are not only willed but determined through rational reflection on the nature of the product. Schelling, as we have noted, disputed the idea that ancient Greek mythology represented a type of primitive natural inquiry into nature in which the gods functioned as abstractions of natural phenomenon. Here, however, mythology progresses towards a *Naturphilosophie* very much like Schelling's in that it establishes a foundation for natural inquiry by imbuing nature with intelligence. Schelling himself thus gestures towards a poetic philosophy with his terminology and his conception of nature, even if his actual prose remains, I feel compelled to add, rather far from eliciting an aesthetic experience.

FREEDOM

If Schelling's explication of organism is to serve as a successful way to evade the dichotomy between materialism and vitalism, it is not enough for it to account for complex natural phenomena such as life and the voluntary motion of animals. It must also account for human will. As throughout his system, Schelling approaches the question of freedom from within nature rather than establishing the two as an oppositional pair. At the point at which Schelling tackles freedom in *The World-Soul*, we see a strange double-restatement that performs a significant reframing of its terms:

> Nature should neither act quintessentially *lawlessly* (as the defenders of a life force have to claim if they are consistent), nor quintessentially *lawfully* (as the chemical physiologists claim), *but rather it should be lawless in its lawfulness and lawful in lawlessness.*
>
> The problem to be solved is therefore this: *how nature can assert an appearance of freedom in its blind lawfulness, and vice versa, how nature can obey a blind lawfulness while appearing to operate freely?* For this unification of freedom and lawfulness we have no other concept than the concept *drive.* (*HKA* VI:216)

Schelling moves from a general statement of the problem and its sought-after resolution in reciprocity in the first paragraph to a restatement that crucially renders the terms unlike by turning freedom into an appearance in the second paragraph. The claimed version of reconciliation of the two in the last paragraph introduces a concept that had also been central to Fichte's attempt to reconcile freedom and nature, namely *the drive.* For Fichte the drive is that which presents itself to the I as a given and which impels activity.[17] All drive is therefore formative drive, since the world and self are formed reciprocally by the positing I that seeks the fulfillment of needs within the world in response to it.[18] For Schelling, the formative drive [Bildungstrieb] is also central; it takes over from the formative force [Bildungskraft] that already organizes inorganic nature geologically (*HKA* VI:252–253). Schelling here not only mentions, but paraphrases, Friedrich Blumenbach when he describes the formative drive "by whose power matter takes on, maintains, and continuously re-creates a particular form" (*HKA* VI:216).[19] For Schelling, the key to understanding the living organism is through the interplay of necessity, that negative force of blind lawfulness inherent to matter, and contingency—or nature's freedom—here observable in the variability of reproduction and healing (*HKA* VI:218). Drive is thus already a complex synthetic force merging necessity and contingency. For Schelling, as Joan Steigerwald explains, the Bildungstrieb "is not the explanatory ground of this union" but rather an expression of a primal interaction between forces.[20] This combination suffuses organic nature in the activities of growth, healing, and generation, and reappears at a higher level in animals, for which it becomes a drive towards movement called *instinct* (*HKA* VI:249). In this form, it manifests a further fusion of irritability and sensibility, and hence a more complex synthesis of necessity and contingency that encompasses the voluntary and involuntary (*HKA* VI:249). Human instinct may be more complex than that of other animals, but it still belongs within this category (*HKA* VI:250). It is instinct that is the "all-vivifying soul" (*HKA* VI:250) of animals.

For Fichte, the drive constituted an experience of a demand that sent the I into the world to sustain its body, which was of the world, thus suturing spirit and world together. The I remained free in its acquiescence to or rejection of the demand, but freedom also inhered in the primacy of the I; Fichte's drive is phenomenological at its foundation and creates the world through its needs. For Schelling, however, the drive is a synthesis of matter and intelligence. Indeed, soul here is itself a synthetic notion; it is instinct; it is accident; it is complexity; it is reciprocal interaction of forces; it is genius; it is free play.

If drive is experienced by all organisms, and instinct by all animals, Schelling by no means denies human exceptionalism. In his *System of Transcendental Idealism* of 1800, Schelling deduces autonomy for the individual in the ambit of practical philosophy. If, in general, Schelling turns from epistemology back to metaphysics, here, extraordinarily, the distinction between the *voluntary* [das Willkürliche], enacted by animals in the *World-Soul*, and the *will* [das Wollen] which humans enact in *The System of Transcendental Idealism*, turns out to be a function of epistemology.[21] Freedom here becomes synonymous with a particular kind of self-consciousness. While Hegel would expand on the notion of self-consciousness as a progressive universal history, it is Schelling who first establishes such a narrative. In the process of increasing self-consciousness intrinsic to Schelling's system, the self first acts. This act makes of the I both a subject and object, but not yet for its own consciousness. The self next becomes conscious of its sensation, that is, of itself as object, and then becomes aware of itself as sensing—and so of itself as subject. "Only in willing [Wollen] . . . the self becomes an object to itself as the whole which it is, that is, as at once both subject and object, or as that which produces" (*System,* 156), that is, that which brings a change into being in the material world. Free activity, in other words, autonomy in the practical or moral sphere, is not the same as merely voluntary activity, which all animals participate in. To will is to want—to envision an ought—in the knowledge that the want can be fulfilled by the self (*System,* 162). The human transcends the animal neither by desiring nor by producing but by understanding the self as the agent of the transition from the former to the latter. Here in *The System of Transcendental Idealism*, Schelling returns to and modifies the concept of *drive.* Formerly a vital impulse common to all living beings and below the level of consciousness or even volition, here the drive is closer to instinct, both free and "spring[ing] immediately and without any reflection from a feeling" (177). The drive is the impulse to bring the objective—the world—into alignment with the ideal—with the ought—hence resolving the contradiction between the idealizing and the intuiting self. Free activity changes intuitions of the world. The productive intuition, however, that is, the perception built upon sensation, follows objective laws. It is only within a conscious self that the concept precedes the object and therefore "free action as such also belongs only to appearance" (*System*, 182). For any human action in the world, in other words, two accounts are equally accurate. One details a sequence of material causes and effects and is objectively true. The second, which entails a goal and a will to bring it about, exists as an appearance in the subject. It is this appearance of goals and will that both "sustains the continuity of self-consciousness" (178) and "make[s] the world become objective to us" (*System*, 182). It is this appearance that is freedom.

To define instinct as spirit, drive as freedom, and will as consciousness of the voluntary is perilously near to inverting the terms that created these dualisms, forcing a radical rethinking of their significance. Under what circumstances could we envision drives and instincts as a form of freedom? If we return to the *World-Soul* essay, we see that Schelling's approach involves widening the arena in which the answer is sought:

> The essence of life, however, consists in no way of a force, but rather of a free play of forces, which are continuously maintained through an outside influence. The necessary in life is the general natural forces which are in play; the contingent, that which maintains the play through its influence, must be a special, or in other words, a material principle.
>
> *Organization and life* in no way express something that endures *in itself*, but only a particular *form* of being, *a commonality of many concurrent causes*. (*HKA* VI:254)

Freedom *in* an organism is the freedom *of* the organization.[22] The forces in play in the most general organization—the world-soul—persist but are supplemented by new forces in play at each level of differentiation and organization, from the inorganic to the organic to the human. While free activity as such is ideal, the free play of forces that manifest themselves as unconscious voluntary activity is real. The term *free play* was most prominently associated with Kant's aesthetic philosophy in the *Critique of Judgment* and Schiller's in *On the Aesthetic Education of Man*. Since Kant contrasts the free play of the faculties that results from aesthetic judgment with both determinate judgments that subordinate experience to concepts or laws (applicable to physical nature) and teleological reflective judgments applicable to living organisms, we need to turn here to the relationship of art to living beings to understand the kind of freedom Schelling's system offers.

THE WORK OF ART

Kant's *Critique of Judgment* is an attempt to make sense of different kinds of judgment elicited by differing relations of intuitions to concepts. Determinate judgments function to subsume an intuition under a concept, to order a particular perceptual experience under a general rule. All simple law-governed phenomena, including physical and chemical sciences, fit into this category as do everyday acts of identification such as "This is a cat." For Kant, the investigation of the workings of living beings, or what he calls organized beings, falls outside this category. Making judgments about the way that organisms function requires accepting purposes beyond the human. Not only do the parts and wholes appear to serve each other reciprocally as ends and means in a living being, but the whole appears to be organized purposively for its own survival. Since only an intellect could conceive of a purpose and design a thing to suit it, the pursuit of the life sciences rests on an "as if" presumption of an intellect behind living forms that would legitimate teleological reasoning. Such a presumption falls outside the realm of evidence and proof, rendering the judgments founded upon them reflective rather than determinate.

Judgments of taste are also reflective judgments for Kant, but while they function in a way analogous to the previous set of judgments, the purpose for which the understanding searches never materializes. While living things must be reflectively accepted as purposive, works of art are purposive without purpose. The appearance of organization in the object under observation spurs the understanding to align intuitions with concepts and organization with a purpose, but both attempts remain without success, tumbling the faculties into an activity that does not come to rest. This is, for Kant, free play. Such free play, however, is not equivalent with freedom of will, which rests not in the realm of judgment but in that of practical reason, of ethics. In fact, Kant, returning to the similar attitude of Socrates in Plato's *Ion*, who denies that artistic inspiration counts as knowledge, is deeply ambivalent about the particular aptitude of the artist that allows her to create works which excite the faculties to free play.[23] Genius is less a capacity than a receptivity, less a skill worthy of admiration than a talent. In other words, not only the observer of art but also its creator acts under a kind of compulsion or as a kind of conduit. Both the creation and admiration of a work of art cause pleasure, but neither requires nor brings about knowledge.

Where Kant theorizes divisions in judgment, however, that create metaphysical gulfs, Schelling's goal is to overcome epistemological barriers by eliminating the metaphysical gulfs that lead to them. Rejecting Kant's "as if," he declares it "an old illusion that organisation and life cannot be explained from natural principles" (*On the World-Soul*, Grant trans. 68) while nonetheless embracing the use of the words "soul" and "spirit." The free play of Kant's aesthetic judgment migrates in Schelling, operating at a number of different levels, most fundamentally within organization itself, and then again also within life, and finally within human endeavors such as the study of nature and art.[24] So, for example, in the *World-Soul*, Schelling claims that the forces described by the dynamic philosophy he is proposing, if taken up by the natural sciences:

> do not in any way serve as *explanations*, but rather only as *limit concepts* for empirical natural science, by means of which the freedom of the latter is not only unthreatened, but is rather secured, because the concept of forces, since each of these admits an infinity of possible degrees, none of which is an absolute . . . opens for it an infinite room for play [Spielraum] within which it can explain all phenomena *empirically*, that is, from the *reciprocity of diverse matter*. (*World-Soul*, Grant trans. 79, trans. modified; *HKA* VI:81–82; see also 216–217)

The freedom of the naturalist here is secured by the reciprocal freedom of the infinite variety of the play of forces within the object of study, which encompasses his own reason.[25] As nature brings forth creatively, but within laws, so can the naturalist exercise the freedom of productive activity within the aforementioned limits. The freedom of the naturalist does not fully mirror the free play of nature, however, just as it does not fully mirror the freedom of the artist. Indeed, the freedom of the naturalist exceeds that of the artist because the naturalist brings forth work in consciousness of self as both subject and object. As we will see, both nature and art, on the other hand, occupy the space of consciousness and unconsciousness simultaneously, which makes them less free but more fully unified.

While Schelling gives us only a hint of his thinking on artwork in *On the World-Soul*, we can already see the way that aesthetics has infiltrated his philosophy of nature in the free play of forces constituting the organism that animates the world.[26] Beginning with *The System of Transcendental Idealism*, this link will become explicit, as he famously declares, "The objective world is simply the original, as yet unconscious, poetry of the spirit; the universal organon of philosophy—and the keystone of its entire arch—is the philosophy of art" (*System*, 12). It is significant that while Kant begins with human cognition, Schelling begins with nature—not only living nature, but nature as a whole—before proceeding to art. For Schelling, the link between art and nature is not merely one of analogy in the way they are cognized, but rather their similarity as products provides evidence of the coincidence of conscious and unconscious activity at all levels of world organization. As Schelling will clearly state in "On the Relationship of the Plastic Arts to Nature," art "occupies the position of an active link between the soul and nature" and has "a productive force similar to [nature]."[27] Leaving aside for a moment the move from living thing to all of nature, we need to follow Schelling's argument about teleology one more time.

For Schelling, nature is produced unconsciously and yet appears as a purposive product. The mechanism by which nature comes to be, in other words, does not merely function blindly like other mechanisms that are propelled forward through cause and effect but coincides with its contradictory model, in which process is drawn forward by means toward an end. The resolution of this contradiction evades both naturalist investigators who imbue matter with a vital essence beyond natural law and those who see matter as entirely inert. Schelling instead posits an original absolute unity that precedes the separation of object from the very concept of concepts, that is, from that which would require consciousness and freedom. A harmony of matter and intelligence eliminates the Kantian gulf between noumena and phenomena and provides a foundation for knowledge of the world. However, it creates a new epistemological problem that demands a new solution. Knowledge of the self requires prima facie a split subjectivity. It should therefore be impossible to become conscious of an absolute identity in which the self participates. Knowledge of this original harmony could only be achieved if (1) conscious and unconscious activity could be recognized in the same intuition (something which happens in the observation of nature) while (2) also in that same intuition, the self would be both conscious for itself and yet also unconscious. This coincidence is a tall order indeed.

Schelling finds a solution in the work of art as a product.[28] The work of art resembles nature in one respect and free activity in a different respect, but goes beyond both. Like nature, the work of art combines the conscious and the unconscious, but in the reverse direction. Nature produces without consciousness or purpose a product that is purposive and hence reflects consciousness (*System*, 219). The artwork, on the other hand, begins as a conscious product, like the free action, and like the free action becomes objective, that is, takes on reality in a material world. Activity only appears free by abolishing the identity of the conscious with the unconscious activity, however. Indeed, all productive activity, artistic creation included, requires a consciousness of the separation of ideal from real (*System*, 221). For this reason, it is only when production stops that an object can appear as a product of self-activity which, however, is not free, as the product, one might say, of inspiration—what one might call with Steigerwald "an unconscious drive into artistic production."[29] We have arrived at the work of art as a unique form of knowledge, produced consciously and yet appearing unfree, mimicking nature in this combination, but at a higher level that reverses the duality.

It is hardly surprising to find that Schelling's description of the unfree element of artistic creation oscillates between the language of *Beglückung* [joy] or *Gunst* [grace][30] (*HKA* IX.I:315) on the one hand, and a dark, uncanny subjection, on the other:

> This unchanging identity, which can never attain to consciousness, and merely radiates back from the product, is for the producer precisely what destiny is for the agent, namely a dark unknown force which supplies the element of completeness or objectivity to the piecework of freedom; and as that power is called destiny, which through our free action realizes, without our knowledge and even against our will, *goals that we did not envisage*, so likewise that incomprehensible agency which supplies objectivity to the conscious, without the cooperation of freedom, and to some extent in opposition to freedom (wherein is eternally dispersed what in this production is united), is denominated by means of the obscure concept of *genius.* (*System*, 222)

The aesthetic judgment entails, on the one hand, the feeling of tranquility that arises from the harmony of consciousness with unconsciousness revealed by the work of art (*System*, 221).[31] On the other hand, the conscious experience of dark, unknown forces that operate against our will hardly sounds like a comforting one. It is a small leap from this description to the character Nathanael's conviction in E.T.A. Hoffmann's short story "The Sandman"—not coincidentally Freud's second touchstone in his essay on the uncanny—that people are "in fact only the horrible plaything of dark powers."[32] Nathanael's impression of lack of agency applies both to life in general and to his own artistic work: "When it was finally completed and he read the poem aloud to himself, he was stricken with fear and a wild horror and he cried out: 'Whose horrible voice is that?' Soon, however, he once more came to understand that it was really nothing more than a very successful poem." Nathanael's oscillation reproduces both the calm satisfaction and the horror of a non-agentic view of aesthetic genius we encounter in Schelling.[33] And for both, the knowledge produced seems to extend beyond the work of art and to illuminate a foundational necessity within human activity. Indeed, the description of destiny [*Schicksal*] would seem to carry us through poetry to mythology. Returning to the language of the later lectures on mythology with which this essay began, we would seem to find here a clear parallel: free will steps into being when the dark forces of destiny are precipitated, not here into mysticism but into a materialist natural science. Genius hence becomes the distinctly uncanny conduit that reveals the unity joining freedom to natural law.

UNCANNY ALL THE WAY DOWN

As we have seen, Schelling is not unique in attributing to genius or to artistic production an element of unconscious activity and, in fact, he draws on a tradition that goes back to Homer and was a commonplace of the Romantic poetic discourse across Europe simultaneous with Schelling's philosophy. Schelling, however, embeds the genius within progressive stages of consciousness, to use Schelling's language from the mythology lectures, which vanquish the unconscious of nature in the appearance of freedom. Art, precisely because it provides access to an otherwise inaccessible primary harmony, reveals the unconscious that should have remained hidden. Art may be the keystone of the entire arch of philosophy, but its very essence is uncanny. The uncanniness of Schelling's system is, however, far more pervasive than merely the conscious and unconscious elements of art production. It lurks on every level of the system and colors all human action. As he explains:

> 1. All action can be understood only through an original unification of freedom and necessity. The proof is that every action, alike of the *individual* and of the entire *species* much be conceived of, *qua* action, as free, but *qua* objective consequence, as standing under natural laws. Subjectively, therefore, for inner appearance, we act, but objectively we never act; it is rather that another acts through us, as it were.
> 2. But now this objective agency, which acts through me, must again be myself. Yet I alone am the conscious, whereas this other is the unconscious. (*System*, 212–213)

We see here the underpinnings of psychoanalysis, but one embedded in metaphysics. The interdependence of conscious and unconscious—their original unity and later opposition—is a foundational element of Schelling's world. It is only in the apparent paring of one from the other that freedom resides. The unconscious spirit that acts through the subject is the intelligence that is coeval and united with matter, "the principle [that] maintains the continuity of the inorganic and organic world, and [that] binds the whole of nature into a universal organism, . . . what the oldest philosophy intuitively [ahnend] welcomed as the common soul of nature" (*HKA* VI:257). We return here to the world-soul, but I want now to stress a different aspect of this animating and organizing principle, namely its reticence: "as the cause of life, [it] veils itself in its own work, withdrawing from every eye" (*HKA* VI:256). The universal organism, we recall, is the play of forces that produces activity and organization throughout organic and inorganic nature; at a higher level of complexity it produces life. By designating the world as a whole an organism, Schelling grants to the world an internally directed, self-animating activity and organization. And by withholding from humans access to a unique force, by granting us only a unique consciousness, Schelling not only has humans partake of this universal organism but encapsulates us within it, as fully of it.

If we return now to the idea that organization of the natural world is created by the free play of forces, we find that Schelling not only relegates human autonomy to a realm of appearance but also subsumes human activity within the free play of an organism larger than it, a being for which that free play, like the free play of Kant's faculties in the aesthetic judgment, is oriented towards its own self, and for which the human participants are therefore mere constituent parts. Schelling insists on the imbrication of the human in a kind of destiny, in a larger unconscious, in a global organism. The unveiling of this reticent, hidden dynamic that is possible only in art therefore renders not only art, and not only the human psyche, but the world itself, uncanny. What emerges from this process is a complex psychology under the sway of an unconscious, an uncanny art, and an ecology that rescues spontaneity at the expense of transparent will.

DECENTERING THE HUMAN

Terry Castle located the Romantic invention of the uncanny in the absorption into the imagination of apparitions and other formerly supernatural phenomena.[34] The psyche thus became the repository of its own unconscious hauntings. At first glance, Schelling's unconscious would seem to instantiate this claim, but a closer look reveals a far more thorough restructuring that ushers us into a house of mirrors. The psyche within which the Schellingean unconscious functions is itself merely a corner of the world in which unconscious intelligence and matter coincide. The Schelling revival that has been in progress for the last two decades sees something other than the uncanny in the absorption of the human into the world's organization, namely the foundation for a philosophy that replaces anthropocentrism with a leveled ontology.

After its beginnings in the 1980s with the work of scholars such as Marie-Luise Heuser-Kessler, Bernd-Olaf Küppers, and Dale Snow, the discovery of Schelling's potential for new theories that not only oppose an instrumental view of nature but undermine a dualistic human-environment paradigm in favor of a single ecological system, has accelerated in the last 10 years.[35] In one of the most creative adaptations of Schelling to a modern nature philosophy, Jason Wirth declares:

> For Schelling, thinking, living, and dying, all of these come to life with *das Wirkenlassen des Grundes*, letting the ground operate (*SW* I/7, 375). This has nothing to do with willing but is, as Schelling felicitously articulates in the 1815 *Ages of the World*, 'the will that wills nothing' (I/8, 235). . . . Only in letting go of oneself and letting the ground operate does the imagination show itself both in human works and as the wild way of nature itself. It is also at the heart of what Schelling will call philosophical religion, the creative—historic yet prophetic—reimagination of the earth and the humans who dwell, often ruinously, on it.[36]

A shift in perspective here dissolves the uncanny by reinterpreting the duality within unity as an opposition without hierarchy. Countering the reification of the subjective that has structured modernity, this realignment integrates the human into nature and undoes the domination inherent in a contrary human will. In this form, Schelling has taken on a new significance in the new millennium.

NOTES

1. Sigmund Freud, "The Uncanny," in *The Standard Edition of the Complete Psychological Works*, trans. and ed. James Strachey (London: Hogarth Press and Institute of Psycho-Analysis, 1955), XVII, 217–256, 224.
2. The translator attributes the piece to Friedrich Hölderlin (1796). Friedrich Hölderlin, "Oldest Programme for a System of German Idealism," in *Classic and Romantic German Aesthetics*, ed. J. M. Bernstein, trans. Stefan Bird-Pollan (Cambridge and New York: Cambridge University Press, 2003), 186–187.
3. Friedrich Wilhelm Joseph Schelling, *System of Transcendental Idealism (1800)*, trans. Peter Heath (Charlottesville: University of Virginia Press, 1978), 232–233. For a reading of Schelling's new mythology in the context of the Anthropocene, see Bruce Matthews, "Schelling in the Anthropocene: A New Mythology of Nature," *Symposium: Canadian Journal of Continental Philosophy* 19, no. 1 (2015): 94–105, 186.
4. Friedrich Wilhelm Joseph Schelling, *Ausgewählte Schriften*, ed. Manfred Frank (Frankfurt am Main: Suhrkamp Verlag, 1985), VI:661. Hereafter referred to as *AWS*. Translations from this edition are mine.
5. See Johannes Fabian's classic *Time and the Other: How Anthropology Makes its Object* (New York: Columbia University Press, 2014). My *Sibling Action: The Genealogical Structure of Modernity* (New York: Columbia University Press, 2017) takes up the European discourse of comparative religion more directly.
6. It is important to note already here that the elimination of dark by light would represent not progress, but annihilation. For more on the unconscious and irrational in Schelling's philosophy of mythology, see Edward Allen Beach, *The Potencies of God(s): Schelling's Philosophy of Mythology* (Albany: SUNY Press, 1994), 47–58.
7. Friedrich Wilhelm Joseph Schelling, *Von der Weltseele—Eine Hypothese der hörern Physik zur Erklärung des Allgemeinen Organismus*, ed. Jörg Jantzen; *Historisch-Kritische Ausgabe*, ed. Hans Michael Baumgartner, Wilhelm Jacobs, Jörg Jantzen, and Hermann Krings (Stuttgart: Frommann-Holzboog, 2000), VI. Citations from this edition will be hereafter

referred to as *HKA*. I will cite English translations of Schelling where available. When citing from German, I will use the *HKA* for works this edition-in-progress has released, turning to other editions where necessary. Translations from the *HKA* edition are mine.

8. Tilottama Rajan discusses the way in which a *Stufenfolge*, or evolutionary series structures nature, history, and mythology for Schelling. "Evolution and its Resistances: Transferences between Disciplines in Schelling's and Hegel's Systems," *Symposium: Canadian Journal of Continental Philosophy* 19, no. 1 (2015): 154–175, 166.
9. See Robert Mitchell, *Experimental Life: Vitalism in Romantic Science and Literature* (Baltimore: Johns Hopkins University Press, 2013), 4–11 for the problems defining the terms and attributing adherents to the sides. The sides crystalize clearly primarily when characterized, or caricatured, by an opponent. For a classic reflection on vitalism and mechanism in the history of science, see Georges Canguilhem, *Knowledge of Life*, intro. Paola Marrati and Todd Meyers, trans. Stefanos Geroulanos and Daniela Ginsburg (New York: Fordham University Press, 2008), 59–97.
10. The move from organism or world, on the one hand, to artwork on the other was not as surprising as it may at first seem, given the long-term understanding of the universe as the creative work of God. See for example, Hume's *Dialogues Concerning Natural Religion*, which lays out standard arguments and compelling critiques of this view.
11. Clara Pinto-Correia, *The Ovary of Eve: Egg and Sperm and Preformation* (Chicago: University of Chicago Press, 1997) and Shirley Roe, *Matter, Life, and Generation: Eighteenth-Century Embryology and the Haller-Wolff Debate* (Cambridge: Cambridge University Press, 1981).
12. See Timothy Lenoir, *The Strategy of Life: Teleology and Mechanics in Nineteenth-Century German Biology* (Chicago and London: University of Chicago Press, 1982), 1–54. Mitchell, as above. While John Zammito has more recently convincingly revised Lenoir's understanding of the relationship between Blumenbach and Kant, the idea of a joining of vitalism and materialism at the root of biology still holds. John Zammito, "The Lenoir Thesis Revisited: Blumenbach and Kant," *Studies in History and Philosophy of Biological and Biomedical Sciences* 43, no. 1 (2012): 120–132.
13. Georg Toepfer, "Organismus," in *Historisches Wörterbuch der Biology. Geschichte und Theorie der biologischen Grundbegriffe* (Stuttgart and Weimar: J. B. Metzler, 2011), I.777–842.
14. No complete translation of *On the World-Soul* exists in English. There is, however, a partial translation by Iain Hamilton Grant from 2010. Translations will be mine when the citation refers to volume VI of the *HKA*. When taken from Grant's translation, I will note that they are his.
15. Friedrich Wilhelm Joseph Schelling, *On the World-Soul*, trans. Iain Grant, in *Collapse: Philosophical Research and Development* 6 (2010): 70 (*HKA* VI:69). For more on the self-organization of nature in Schelling, see Marie-Luise Heuser-Kessler, *Die Produkitvität der Natur. Schellings Naturphilosophie und das neue Paradigma der Selbstorganisation in den Naturwissenschaften* (Berlin: Duncker und Humblot, 1986), 29–53.
16. For more on Schelling's *Timaeus* commentary in the context of nature and the divine, see Bruce Matthews, *Schelling's Organic Form of Philosophy: Life as the Schema of Freedom* (Albany: SUNY Press, 2011), 103–133.
17. Johann Gottlieb Fichte, *Das System der Sittenlehre, Gesamtausgabe der Bayerischen Akademie der Wissenschaften*, ed. Reinhard Lauth and Hans Jacob (Stuttgart-Bad Cannstatt: Friedrich Frommann Verlag, 1965), V, 125–143.
18. For more on Fichte and the drives, see my "Love or Knowledge: Sexual Epistemology in Fichte and Kleist," *Germanic Review* 92, no. 4 (Fall 2017): 368–387.
19. Schelling: "kraft dessen sie [die Materie] eine bestimmte Gestalt annehme, erhalte, und immerfort wiederherstelle"; Blumenbach: "ein besonderer, eingebohrner, Lebenslang thätiger würksamer Trieb liegt, ihre bestimmte Gestalt anfangs anzunehmen, dann zu erhalten, und wenn sie ja zerstört worden, wo möglich wieder herzustellen." Johann Friedrich

Blumenbach, *Über den Bildungstrieb und das Zeugungsgeschäfte* (Göttingen: Johann Christian Dietereich, 1781), 12.

20. Joan Steigerwald, "Epistemologies of Rupture: The Problem of Nature in Schelling's Philosophy," *Studies in Romanticism* 41, no. 4 (2002): 545–584, 555.
21. This distinction becomes more complicated when Schelling grants *Wollen* or willing-of-itself (241) to the world in his 1806 "Treatise on the Relationship of the Real and the Ideal in Nature," trans. Dale Snow, *International Philosophical Quarterly* 55, no. 2 (2015): 235–250. In his 1809 *Philosophical Investigations into the Essence of Human Freedom*, Schelling distinguishes human will from a will of nature, "will that beholds itself in complete freedom, being no longer an instrument of the productive universal will [schaffenden Universalwillens] in nature, but rather above and outside of all nature." Friedrich Wilhelm Joseph Schelling, *Philosophical Investigations into the Essence of Human Freedom*, trans. Jeff Love and Johannes Schmidt (Albany: SUNY Press, 2006), 33. *Philosophische Untersuchungen über das Wesen der menschlichen Freiheit und die damit zusammenhängenden Gegenstände*, ed. Thomas Buchheim (Hamburg: Felix Meinter Verlag, 1997), 36.
22. As Lara Ostaric puts it, "the principle of life is an expression of nature's freedom." "The Concept of Life in Early Schelling," in *Interpreting Schelling: Critical Essays*, ed. Lara Ostaric (Cambridge: Cambridge University Press, 2014), 48–70, 62.
23. Plato, *Ion*, trans. Paul Woodruff, in *Complete Works*, ed. John M. Cooper (Indianapolis and Cambridge: Hackett Publishing Co., 1997), 937–949. Immanuel Kant, *Critique of Judgment*, trans. Werner Pluhar (Indianapolis and Cambridge: Hackett Publishing Co., 1987), 174–178 (AA 307–310).
24. See Devin Zane Shaw, *Freedom and Nature in Schelling's Philosophy of Art* (London and New York: Continuum, 2010), 63–87.
25. As Leif Weatherby puts it, "the reciprocal causality of the organism mirrors the form of judgment" in *Transplanting the Metaphysical Organ: German Romanticism between Leibniz and Marx* (New York: Fordham University Press, 2016), 177. See also Lara Ostaric, 57–58.
26. See also Gabriel Trop, "The Aesthetics of Schelling's Naturphilsophie," *Symposium: Canadian Journal of Continental Philosophy* 19, no. 1 (2015): 140–152.
27. Friedrich Wilhelm Joseph Schelling, "Concerning the Relation of the Plastic Arts to Nature," trans. Michael Bullock, appendix to Herbert Read, *The True Voice of Feeling: Studies in English Romantic Poetry* (London: Faber and Faber, 1953), 321–364, 324.
28. As Paul Guyer notes, "Schelling transformed Kant's conception of aesthetic ideas as a form of free play with truth back into a more traditional conception of an apprehension of truth." Paul Guyer, "Knowledge and Pleasure in the Aesthetics of Schelling," in *Interpreting Schelling: Critical Essays*, ed. Lara Ostaric (Cambridge: Cambridge University Press, 2014), 71–90, 71.
29. Steigerwald, "Epistemologies of Rupture," 569.
30. Guyer, "Knowledge and Pleasure," 83, evaluates this pleasure as a negative one arising from the cessation of pain caused by contradiction inherent in "the fundamental paradox of metaphysics." While Guyer is correct that Schelling has moved away from the self-activity of Kant's state of free play as a cause of pleasure, Schelling goes out of his way rhetorically to add the affects of astonishment and joy (*System*, 221) as positive emotions beyond a simple satisfaction in resolution.
31. One might equally perceive in this description of an identity that cannot be perceived in the producing, but only in the product (*System*, 212–223), the precursor of the Marxian commodity fetish that unites in itself what is distributed among the laborers that create it, and hence siphons their will and agency.
32. "The Sandman," *Tales of E.T.A. Hoffmann*, ed. and trans. Leonard Kent and Elizabeth Knight (Chicago and London: University of Chicago Press, 1972), 137–167, 107, 109.

33. One might in fact recognize in Hoffmann's figure of Coppola/Coppelius, who personifies the dark powers of fate and their imbrication in the powers of nature, an echo of Schelling's *copula*. Schelling uses the word *copula* in his essay on the "Relationship of the Real and the Ideal in Nature" to refer to the bond between the finite and the infinite, the "individual in the whole and the whole in the individual" (246), a bond (although not a strict equation) that encompasses opposition and identity to enable all complex organisms (in the sense we have been using the word here), and hence also life.
34. Terry Castle, *The Female Thermometer: Eighteenth-Century Culture and the Invention of the Uncanny* (New York and Oxford: Oxford University Press, 1995), 168–189.
35. Tilottama Rajan, "Evolution and its Resistances," 156–157, uses Bruno Latour's definition of ecology to explicate this element of Schelling's thinking. Bernd-Olaf Küppers, *Natur als Organismus: Schellings frühe Naturphilosophie und ihre Bedeutung für die moderne Biologie* (Frankfurt: Klostermann, 1992). Dale Snow, *Schelling and the End of Idealism* (Albany: SUNY Press, 1996), 67–118. For more recent instantiations, see Bruce Matthews; Iain Hamilton Grant, *Philosophies of Nature after Schelling* (London and New York: Continuum, 2006); and Jason Wirth, *Schelling's Practice of the Wild: Time, Art, Imagination* (Albany: SUNY Press, 2015).
36. Wirth, *Schelling's Practice of the Wild*, 143.

9

"IT . . . LIVES BY DYING"

S. T. Coleridge's Non-Vital Life and Colonial "Necoral-Politics"

Lenora Hanson

NECROPOLITICS AND ROMANTIC LIFE

Achille Mbembe has defined the concept of necropolitics as "the creation of *death-worlds* . . . in which vast populations are subjected to conditions of life conferring upon them the status of '*living dead*.'"[1] Mbembe discusses necropolitics primarily as a contemporary political phenomenon but loosely refers the concept back to the Romantic period through Hegel and the colonial slave trade.[2] Here I want to argue for deeper entwinement between Romanticism and necropolitics through the chiasmic structure—evoked in the figures of "*death-worlds*" and "*living dead*"—Mbembe gives to it.

Indeed, this interpenetration between opposing states, here of life and death, is a central feature of Romantic physiology and political economy. Consider, for instance, the physician John Brown, who reconciled sickness and health, excitation and languor, even life and death not as opposing forces but as variations of the same force. As Paul Youngquist writes, Brown understood these variable states to be "effects of the same interplay of forces" in a monistic physiology in which "what menaces life is life itself."[3] Or consider Catherine Gallagher's argument that, in the Romantic era, national economic growth and vitality was understood to be dependent upon the *expense* of the life of laboring populations. Like Brown's physiology, Romantic political economy was an "autotelic system, not only tolerating, but also requiring dynamic conflict."[4] In short, Romantic forces of life and death—far from being dialectical or oppositional in a straightforward sense—were thus often internalized in a chiasmus like that suggested in the "*death-worlds*" and "*living dead*" of necropolitics. Whether in physiological or economic terms, life and death were often put into a relation of interchangeability and indistinction rather than of binaries or othering.

Such a relation offers a figure of indistinction rather than one of exclusion. Rather than read the nonliving as a site of unease, I argue that the ambiguity between the living and the nonliving—conceived of as both human and nonhuman forms—was far more important as a productive force within the development of a liberal mode of global sovereignty in the Romantic period. If the specter of the nonliving or the automated haunted Romantic notions of life, as essays in this collection demonstrate, then I am suggesting that this haunting was not so much disruptive as it was productive of an emerging mode of sovereignty that demanded bodies and populations that could be rendered obsolete or nonessential. Drawing from recent work on biopolitics and critical race studies, I approach the central theme of this collection as a way to grapple with the ambivalent inclusion, or what Daniel Nemser has called the "*differential exteriorization,*" of otherness within a global hierarchy of life.[5]

Here I consider the chiasmic relation of Mbembe's "*living dead*" as a way to rethink the politics of figuration in the Romantic period. Romantic scholars have oftentimes turned to figuration, chiasmic and otherwise, in order to grasp language in its noncognitive and nonprogressive potentiality. Thus, in their introduction to *Dark Romanticism*, Jacques Khalip and Forest Pyle have recently linked figuration to a kind of organismic and nonhuman perception. As opposed to what is clearly visible or articulated, figuration is an "almost proprioceptive interest in what *is there* in its shadows as opposed to something that emerges out of its negation or utter deletion." For them, the "non-knowledge of tropes" offers an alternative to knowledge defined as what can be brought fully into view. Figuration harbors alterity within its always indistinct configuration of language and the world. Relatedly, Tilottama Rajan has argued that Romantic notions of organic nature remain relevant for us today because they demonstrate the "*supplementary* recourse" by which "beauty [was] *troped* as natural." That is, for Rajan, and perhaps for Khalip and Pyle too, the supplementary construction of Romantic life "might actually be used to deconstruct . . . the relation between 'life' and 'organization,' the different kinds of animate organization, and the location of the border between organic and inorganic, human and inhuman."[6] In the former, Romantic figuration offers a latent alterity to history, in the latter a rhizomatic nature of nature.

Without doing away entirely with figuration's potential, this essay considers it as a reference for a different kind of dark Romanticism than the one Khalip and Pyle propose.[7] Instead, the borderlessness between life and nonlife might best be characterized as a necrotic intimacy that was distributed and attributed to certain bodies and territories in the Romantic period. Such intimacy did destabilize clear divisions between opposing states of life and death, or human and nonhuman, as Rajan suggests. But, as Mbembe and other scholars of the biopolitical have argued, this destabilization was consolidated within particular bodies.[8] In other words, it was a controlled contamination. The porous border between the vital and the necrotic became a way to divide certain bodies and zones from others, or to unevenly allocate the fluid condition between life and death. Uneven distribution has long characterized our understanding of the twentieth century colonial world, which Frantz Fanon described as "a world divided in two." Less attention has been paid to the more chiasmic—and Romantic—mode of that division, to which Fanon gestures when he writes that colonialism is both a "geographical *configuration and classification*."[9] Like Mbembe's chiasmic "*death-worlds*," Fanon forces division and relation together, in a figurative logic that resists any easy or clear understanding. This conjuncture of division and relation returns in Mbembe's notion of race, which he describes as "a doubling, an uncertainty, and equivocation . . . a fatally imperfect language, grey and inadequate." Between Mbembe and Fanon, we can say that the uncertainty upon which racial and colonial classification operates is not a site of complete openness or porosity, but of an uneven distribution of porous states to certain bodies and populations. Such division relies on an unstable relation of the kind figuration affords. This is, in part, what Mbembe suggests when he describes necropolitics as a power that "takes control of and vests itself in . . . a split between the living and the dead . . . [that] defines itself *in relation to* a biological field."[10] While Mbembe and Fanon more often emphasize the splitting function of power, it is the problem of vesting, configuration, and relationality as it accompanies division that I interrogate here.

This notion of power as a vesting within, in order to split from the inside, aptly captures the rhetoric of Coleridge's 1816 "Theory of Life." The liberal logic he applies to the concept of life (he even imparts to metal what he calls "vital essence"), as I argue here, is deeply enmeshed with the necropolitical history of British colonialism. I treat Coleridge's incorporation of "the lifeless" and "physical and dead properties" into a general system of life alongside James Stephen's 1802 *Crisis of the Sugar Colonies* and its recommendation for a "more liberal policy" of "a new and happy system of colonization."[11] Supplementing a critical tendency to err on the side of reading alterity or porosity within Romantic-era notions of life, I instead consider how Coleridge's undoing of binaries in his 1816 "Theory" reproduces a colonial response to the threat of slave rebellions in 1802. As I argue here, a necrotic intimacy between life and death mutually structures both Coleridge's rhetoric of species classification and Stephen's liberal colonialism in the early nineteenth century.

"THEORY OF LIFE" AND "NECORAL-POLITICS"

Coleridge's "Theory of Life" was meant to intervene in the vitalist debates raging in 1816, which centered on the methodological differences of describing a *vis vitae*, or principle of life, and the extent to which the science of life could unify all of science.[12] Although it went unpublished until the mid-nineteenth century, "Theory" responded by rejecting any division between "things with life" and "things without life", instead "generaliz[ing] the idea of Life . . . more broadly" to make "those physical or dead properties . . . themselves vital in essence."[13] Jacob Risinger has argued recently that "in refusing to create distinctions and thresholds that would make it possible to define life in terms of its absence—Coleridge dismissed a coercive model of biopolitics that Michel Foucault would later characterize as one that called an individual 'existence as a living being into question.'"[14] But it is this critical tendency to embrace indistinction as a term of openness and inclusivity in Romantic conceptualizations of life that I query here. That is, I want to consider how Coleridge's radical incorporation of the "physical or dead" into vitalism ultimately divides the mechanism from the purpose, or means from the end, of life in "Theory."

In a fragment written a few years after "Theory," Coleridge evocatively renders the indistinction to which Risinger refers. But his "refus[al] to create distinctions" does not operate throughout his system of life there. Rather, it is consolidated within a specific form of life. In that fragment Coleridge includes an extensive footnote to describe the corallighine slime, a species that, like the polyp, was thought to be both plant and animal. Interested in the ossifying decay of its reproduction, he writes that "[i]t grows . . . even as gristle becomes bone,—and thus we may truly say, lives by dying."[15] Coleridge's interest in these skeletons of life, an interest sustained at least between 1816 and the writing of this fragment in 1822, is that they seem to materialize or objectify species reproduction in general. In the coral he sees what Rajan describes as the "synonym for generation" for which reproduction stood in the early nineteenth century.[16] Reproduction, in her words, denotes a "metaprocess" in which "the species takes in what is different, yet continues the same."[17] Coleridge's figuration of the coral as living by dying suggests that they embody this "metaprocess" by which reproduction turns difference into sameness. The generational sense of reproduction thus becomes an objective or concrete feature of the coral, where a drama of species plays out in real time. In "Theory" he elaborates upon this reproductive process as follows: "In the CORALS and CONCHYLIA, *the whole act and purpose of their existence seems to be that of connecting the animal with the inorganic world by the perpetual formation of calcerous earth*. For the corals are nothing but polypi, which are characterized by still passing away and dissolving into the earth, *which they have previously excreted*, as if they were the first feeble effort of detachment."[18]

In contrast to other classificatory schemas that divided living from nonliving things, Coleridge centers his system on a species whose very existence and purpose is "that of connection." His coral are ever-translating engines that put opposing states into relation. Where Bichat and others define life through function, and thus make distinctions between living and nonliving things, Coleridge produces a form whose very function is to collapse the borders between dead matter and living form. His coral configure the work of classification, forging relations in the "perpetual formation" of a natural history of which they are the "first feeble effort" of what Coleridge later calls "individuality . . . in its first dawn."[19] But this translational existence is not entirely without residue, because along with perpetual formation Coleridge also defines their existence as a perpetual excretion. The materiality of the coral—their physiological manner of reproduction—is both pure connection, what is without excess, and what is, by definition, an excess that can be erased in its "passing away and dissolving into the earth."

Coleridge's entire "Theory" depends on the inclusion of these living dead into his system of life. Such radical incorporation provides the ground by which to critique other theories that are incapable of unifying death, or those "heterogeneous and adverse powers," with the self-renewing processes of life. Buffon and others give a "definition of Life, as consisting in anti-putrescence, or the power of resisting putrefaction," but this "substitutes an abstract term [and] presents to us no *inclusive* form out of which the other forms may be developed."[20] Coleridge's account, in contrast, accommodates the negation of life as part of the vital process "out of which the other forms may be developed."[21] Thus, he argues that such forms must be included if life is to be understood as a "unity in multeity" rather than reduced to a mechanical arrangement of parts, if it is to constitute "a real existence" whose end is vital rather than functional. Such life, like the imagination, demonstrates "the principle of unity in *multeity*" without abandoning an ascending hierarchy that develops from the "utmost *latency* in which life is one with *the elementary powers of mechanism* . . . to its highest manifestation [of] *life as life* [into] an ascending series of intermediate classes and of analogous gradations in each class."[22]

The manner in which coral, conchylia, and polypi are included in Coleridge's system is a curious one, then, in that it renders them simultaneously central to and expendable from the ascending reproduction of "*life as life*." Coleridge's "inclusive form" of such life blurs the boundary between mechanical inorganic parts and the vital unity of life while retaining the former as a developmental phase of the latter that can be excreted or dissolved in the process of reproduction. Coral continuously undo oppositions, forging what Rajan calls "strange hybrids . . . between mechanical and organic," between "organic and inorganic, human and inhuman." Indeed, this species puts figuration's capacity for "transformation, substitution, or exchange" into living form, producing an uncanny encounter between language and nonhuman life in coral's manner of reproduction.[23] But the medi-acy between life and nonlife in these objects is not supplementary or rhizomatic, and its does not spread out to contaminate other forms. Rather, the intimacy between mechanical nonlife and vital life is consolidated in the coral and its kin species. Coral's connection, or configuration, of unlike things becomes the base and lowest part of Coleridge's classificatory order. We might say, in Mbembe's parlance, that within Coleridge's anthropocentric natural history, the coral become their own "*death-worlds*" in the act of their very reproduction. The "act and purpose" that brings them into the generalized idea of "life as life" maintains, rather than abolishes, the mechanical feature that makes them both necessary and expendable. The necrotic intimacy that enables them to establish an economy between the nonliving and the living is the very feature that signifies their superfluity in the last instance. Tasked as objects of classification *and* forms of flux, as a stabilizing indistinction within a classificatory hierarchy, the coral do the dirty work of vital existence.

And, at the end, their defining indistinction becomes the point of distinction between an economy of substitution and the "highest manifestation [of] *life as life*," as the sovereign form of Man. If coral are defined by their figurative ability to substitute or exchange, then Man is defined by an individuality that turns such exchange into a surplus. The conclusion of "Theory" does not waver on this point: "Man . . . has the whole world in counterpoint to him, but he contains an entire world in himself . . . the apex of the living pyramid . . . he is referred to himself, delivered up to his own charge: and he who stands the most on himself, and stands the firmest, is the truest, because the most individual, Man."[24] Coleridge's inclusion of the mechanical within the vital, the nonliving within the living, is thus the liberal means to an absolute end. The latter stands outside of the economy of inclusion by way of containing it, perfecting it. Man is the container of a world from which he stands apart as the perfect scale, embodying autonomy rather than mixing distinctions. Coleridge thus brings inorganic, mechanical life into the "vital agency" of what he calls "life as life," in which a sole transcendent difference is preserved.

His treatment of the coral as mechanically vital produces a subtle rupture in his account of nature, between its descriptive or phenomenal and its symbolic or teleological mode.[25] The coral's epigenetic qualities of hybridity are made into a mechanistic excess of life, which passes away in the realization of a primal distinction between human and nonhuman. In other words, the connective function that brings "calcerous earth" and the "inorganic world" into vital life is simultaneously instrumentalized. Coleridge instrumentalizes indistinction's blurring of binaries, rendering the "very lowest species" separable from the purposive arc of species and its "law of vital action."[26] Like his other attempts to produce the Symbol as the immediate presence of a universal truth, Coleridge's figurations of nonhuman life here lead, in Ian Balfour's words, to a "regime of hierarchy and exclusion."[27]

If this reading seems to be overburdening the coral, then it will be helpful to note that Coleridge understood "the most pregnant historic Symbol of Earth [to be] a Coral Bank on a Stratum on Coal."[28] As Symbol these depictions are *meant* to exceed themselves, such that CORALS and CONCHYLIA become ambiguous participants in a philosophy of history, living not only a discrete biological life but also that of a universalizing telos. These "lowest species" ensure the unfolding of progressive history, materializing the universality of vital life in order to be transcended by it.

Such indistinction is posited not simply as an epistemological limit of science but as, truly, essential. Coleridge's Symbols transcend a purely epistemological or classificatory regime in favor of an ontological revelation: "In the polypi, corallines, &c., individuality is in its first dawn; there is the same shape in them all, and a multitude of animals form, as it were, a common animal. And as the individuals run into each other, so do the different genera. *They likewise pass into each other so indistinguishably, that the whole order forms a very network*."[29] The reproductive commonality of these "multitude of animals" grounds Coleridge's classificatory schema, implying that the taxonomic failure to divide species properly is not a linguistic or scientific limit but an accurate expression of the indistinguishable nature of life "in its first dawn." These lowest species function as the irreducible and common form within a "generalize[d] idea of Life." They are the specimens through which Coleridge can claim that "physical or dead properties are themselves *vital in essence*."[30] But as we see above, Coleridge brings the mechanical or nonliving into vital essence in a way that renders it superfluous. That is, there is life as "vital essence," which extends even to "the life of metals," and then there is the "life as life" that contains but stands apart from it in the "primary and universal form of vital agency" of Man.[31]

Such inclusion organizes everything—death, metal, inorganic matter—under the banner of life, extending to each thing a role in its self-renewing process. Included within the generality of "life as life," a version of mere life, these forms exist for the sake of perpetual death. Coleridge's incorporation of the nonliving or non-vital within his more generalizing idea of life relies upon life's figurative potential to connect unlike things but turns such potential into a primal phase of physiological individuation. Such disregard takes what is common to life, the network or connection of ecological interdependence, as the nebulous "first dawn" of individuation and distinguishability. It takes what is most common and translates it into what is most base, pushing physiological function down to the bottom of a hierarchy of forms.[32] This inclusion operates through what Mbembe described as the "doubling . . . uncertainty . . . and equivocation" of race, allocating it not to human bodies but to the lowest of species in Coleridge's vitalist taxonomy.

Thus, Coleridge's "Theory" is littered with the parts that are cast off and excreted in the realization of the ascending series of life. Here his comments on the shell, a feature used to identify forms akin to coral, are illustrative. Coleridge resolutely rejects the shell as a *vital* organ that could be used to classify testaceous animals. A feature of function rather than essence, it is superficial and could never be "co-extensive with the organized creation" or symbolize the *vis vitae*.[33] Nonetheless, despite Coleridge's protestations against it as a vital organ, his inclusive taxonomy hinges on it. The shell is, in fact, central to his notion of the coral's connective reproduction and its "*perpetual formation of calcerous earth*," which acts as the transitional and transformational matter of life. Calcareous earth *was* recognized as a kind of shell in the Romantic era, considered to be a limestone that was "supposed to have been formed entirely from shell or bone."[34] This earth, which Coleridge claims as the abject engine of vital incorporation, was, as Erasmus Darwin described it, "the accumulated exuviae of shell-fish," the exuviae that the *Oxford English Dictionary* defines as the "cast skins, shells, or coverings of animals; any parts of animals which are shed or cast off."[35] Such superficiality, what can be like a garment stripped, or from a body divested, of what can be skinned from living form, is constitutive of Coleridge's economy of life. This exteriority provides the unstable ground for Coleridge's incorporating economy, which, as I have already suggested, cannot do without the production of material debris. Such incorporation spatializes the relation between inside and outside, through the designation of what can be cast off and left aside *from within* a living system. Coleridge inadvertently hinges his theory of a universal or general "life as life" upon the accumulation of a non-vital excess. His "most pregnant historic Symbol[s] of Earth" quickly invert to become more like Mbembe's "death worlds," a species defined by the contamination of life and death.

This specter of excess is rendered elsewhere in "Theory" as a relic or residue. The "lowest species" of his taxonomy are "*relics of the lowest orders*" or "residue . . . of vegetable and animal life."[36] Alongside these calcareous remains, various fungi are associated with "*the dark recesses* of the mine" and an "almost *shapeless*" form.[37] Together, these references conjure a primordial fantasy of the nonhistorical, of what is eternally preserved outside of time as the "prehistorical nature [of] pure timelessness" that Benjamin read into the pastoral.[38] Like the process of interiorization, Coleridge's figuration of reproduction wants to be timeless, without excess. Relics and residue are not metonymic for him, which would risk a nonidentity between part and whole. Internalized within the perpetual formation of progress, he wants them to impart complete continuity between the timeless and the present. Hence his declaration at the beginning: that what he is after is not a "history [but] the law of the thing."[39]

But although Coleridge seeks to abolish history, "Theory" is nonetheless marked by the historical. The calcareerous excess and residual relics that cling to the coral appear like a return of the repressed, imparting to "Theory's" account of life a more fragmentary and transient quality. These gestures, as I suggested above, disclose a violent gesture at the center of Coleridge's theory, one in which certain forms of life must be instrumentalized and expendable in the production of "the highest manifestation [of] *life as life* [as] an ascending series."[40] But this gesture also exposes the fault lines of any such unifying account of life, which as we see in Coleridge paradoxically grounds itself on the life—the coral—that is most expendable from it in the last instance. Both the crux and the excess of Coleridgean life, the coral are more than the name of a nonhuman species. They also index an irreducible aspect of the historical, in a Benjaminian sense, as a violence that cannot be erased, or that cannot be wholly incorporated into the progressive development of time from its "first feeble efforts" to the "highest manifestation [of] *life as life*." In short, while Coleridge treats them otherwise, his coral, conchylia, and polypi assume an allegorical function within his symbology. Although Coleridge relies on the coral as a species defined by equivocation (Are they life or death, plant or animal?), their status as shell and bone, relic and residue implies an altogether different quality of the uncertainty of natural history. In this sense, we can read Coleridge's acknowledgment that the time of the coral is one of "perpetually . . . passing away" as an unconscious reminder of what Benjamin called "the experience of eternal transience . . ." in history, but as one in which what passes is never fully past.[41] The coral's exogenous superficiality, its accumulated remains, offers a more allegorical temporality within one that naturalizes incorporation without a trace. This analogy between Coleridge's coral and Benjamin's ruins suggests that we must look to the means of its production, so to speak. What forces, we might ask, compressed the coral into such allegories of life and death, of interiority and excess, of chiasmus and taxonomy?

Here I turn to the context of colonial reconfiguration and slave rebellion to argue that Coleridge's species logic bears a striking resemblance to the liberal discourse about slave bodies and their place in global, historical progress.[42] Much like Coleridge's "Theory," that discourse sought to create what Gavin Walker has described as the paradox in which "hierarchy [is] a form of equivalence or . . . commensurability."[43] It is in this sense that the coral do become figures of history. Not as a "unity in multeity" or as rhizomatic indistinctions but as dialectical images pulsing with the decay and collision of history, which is to say the shattering possibilities of history's ruins and exteriority. In this final section I move to consider the terms of interior and exterior, of vital and non-vital, that we have seen in Coleridge's "Theory" as they were deployed in a recommendation by prominent abolitionist James Stephen for a more liberal model of colonialism, and in response to the Haitian Revolution and contemporaneous slave rebellions in the West Indies.

GLOBALIZING LIFE AND COLONIAL REBELLION

When at the end of "Theory," Coleridge invokes Man as Symbol of the equilibrium attained between law and Nature, he does so atop the layers of exoskeletons, gristle, and shell of other forms. In this concluding section, I suggest a figurative sympathy between Coleridge's forms that "live by dying," Mbembe's contemporary "*living dead*," and the colonial fascination with the physiology of slave rebellions.[44] I do not intend to charge Coleridge's "Theory" with a repressed colonialist gaze as much as I wish to show how the figurative work of "transformation, substitution, or exchange" of the vital and non-vital in his system of life uncannily reproduces that logic of configuration that Fanon attached to colonialism. This is to say, Coleridge's use of the concept of life is as inseparable from a racialized and colonial history as its contemporary use in biopolitics.[45]

In this final section I turn briefly to James Stephen's 1802 *Crisis of the Sugar Colonies*, published amidst the French campaign to retake Santo Domingo from Toussaint L'Ouverture. In that text, the chiasmus of internal and external, and of life and death, are rendered in similar and physiological terms as in "Theory." Far from the cosmopolitan "horizontality, transversality, or mutually constitutive nature of the relations between peoples and nations" Evan Gottlieb has recently detailed in Romantic globalization, that chiasmus underpins a "more liberal policy" in the spirit of a "new and happy system of colonization."[46] I want to emphasize that Stephen's text is explicitly abolitionist, in keeping with his lifelong position, and that his recommendations are made in an attempt to "meliorate[e] the condition of the great mass of the people" enslaved in the West Indies. Even so, what I am interested in here is the way that this "liberal policy" is supported by the claim that this new arrangement will not alter the essence of empire but rather preserve it. Indeed, emancipation is said to be the "ground of necessity" that will maintain British global hegemony. This ground weaves the bodies of slaves into the necrotic intimacy of liberal universality already seen above, but here spatialized within global empire.[47]

Put in the terms of "Theory," the general idea of empire necessitates both connection and exclusion, an inclusion of parts that can, in the last instance, be rendered separably non-vital. Thus, Stephen advocates for the incorporation of slave populations into the British empire as a transitional population that can be defined as neither slave nor laborer.[48] Stephen argues in favor of slavery reform through a quasi-organicist argument about the British empire, in which its global balance of power, or hegemony, depends upon its capacity to dynamically and progressively reproduce—or, in Rajan's words, to "tak[e] in what is different, yet continu[e] the same."[49] Central to such incorporation, however, is an uneven global physio-geography that identifies the West Indies with death and pestilence and slave bodies with the fluidity and flux of nonsovereign bodies. Like Coleridge's coral, they are configured as both central to and superfluous in global political harmony. Stephen draws physiology and geography into a necrotic assemblage of the British empire, speaking the language of organic harmony and inclusion that would later show up in Coleridge's intervention into the vitalist debates.

That the West Indies are a geographic zone of indistinction is made clear by the flux between freedom and slavery outlined in Stephen's text. He is at once at pains to clarify the freed condition of former slaves in Guadeloupe, Cayenne, and Santo Domingo, in the wake of rebellions there, and in order to prove that the French expedition there is an attempt to re-enslave those populations, he writes: "a great majority of the public, being ignorant . . . of the distinguishing character of negro-bondage, is of course liable to much imposition and mistake in judgment of these revolutions by which that bondage has been abolished, and of the important changes which have been produced . . . which may be of dangerous consequence."[50] This definition of the "essential difference" between the "true nature" of slave bondage and the current freedom in Guadeloupe and Cayenne is forged in the context of geopolitical crisis and what for him is a historical transformation from which there is no going back. The pains of classification he takes are for the sake of a definitive distinction between slavery and freedom, such that readers understand "with precision, the true nature of that condition inadequately defined by the term *slavery* from which the French negroes have passed, and its *essential difference* from that to which they have attained."[51] That true nature of slavery is privatized ownership and punishment, in contrast to the public market of labor and court of punishment under which citizens live. It is this latter condition of "essential difference" into which the freed slaves of Guadeloupe and Cayenne, but importantly not Santo Domingo, have passed, a fact of which "a great majority of the public" is ignorant.

But this essential difference turns out to be relative. At the same time that Stephen asserts this difference to be essential, he also qualifies it in relation to

English laborers. The true nature of freed slaves hovers between positions, a point first clarified in Stephen's assertion that there is no proper name for a freed slave: "There are no proper and peculiar names to distinguish the state of the negro in bondage, from his enfranchised condition."[52] A freed slave is not the same thing as a freeman, but neither is he a slave. Thus, the term freeman, when applied to a slave, is an imperfect translation that cannot be fully equated with an English laborer. Instead, the freed slaves of Guadeloupe and Cayenne remain in a state of negative definition, of linguistic lack. The difference between slave bondage and the current condition of "the French negroes" is not, it turns out, identical to freedom, but rather identical to transition and indeterminacy.

Extending the lack of language into an account of history, Stephen makes the transitional non-name of freed slaves into a mark of their developmental indistinction: "Look at the most laborious peasant in Europe, and if you please, the most oppressed: he is toiling it is true from painful necessity; but it is necessity of a moral kind, acting upon his rational nature."[53] But "[w]hen you talk to [the negro] of the rewards of industry, and the evil consequences of indolence, you speak a language he can but imperfectly understand. Hopes and distant fears, as incentives to work, are to him as a new science whereof he has the very elements to learn; or rather like senses, the organs of which are become from want of use inflexible and unsusceptible."[54] Although Stephen puts the freed slave definitively outside the relation of dominance that characterizes slave-bondage, this does not mean that he has attained the freedom of individual labor: "the self-dependency of a rational being, the close connection between his conduct and his natural, or social welfare, are ideas perfectly new."[55] Thus, the fact that freed slaves have no proper name is yet another way of registering their indeterminate or indistinct position within history, which is above registered by the lag between culture and nature. This impropriety of namelessness, and the nonidentical relation with the English laborer to which it attests, indexes historical progress between Europe and the West Indies. This position emerges most clearly in Stephen's claim that there has not yet been a real struggle for freedom inaugurated by slaves. Both name and revolution are permanently to-come for the slave.

The "essential difference" of the freed slaves is really, then, an indeterminate condition determined in relation to "a rational being." Intermingled between slave and laborer, they figure the thresholds of global development. This way of putting difference into an uneven relation of equivalence—free but not rational—underlines the chiasmic nature of sovereignty in which no contradiction persists between a spatialization of what Mbembe qualifies as "full subjects capable of self-understanding, self-consciousness, and self-representation" in one place and of nonsubjects elsewhere. This geographical distribution maintains the uneven development of revolutionary history, in which slave revolutions remain imprecise and indistinct, not-yet existent within the Western political "project of autonomy . . . communication and recognition."[56]

The unevenness of this historico-geographical distribution, designated by naming, finds its physiological ground in the racialization of vulnerability and porosity that was assigned to the bodies of slaves and indigenous bodies in eighteenth- and nineteenth-century science. This vulnerability put slave bodies into association, or in Fanon's terms, in configuration, with nature and the nonhuman.[57] And like the necrotic intimacy of Coleridge's coral, it brought life and death together in the same form. Thus, as in countless other Romantic-era texts, Stephen asserts that "disease and death ever have been, and ever will be found, marching in the train of a West India army, and cutting down its battalions with great and deplorable rapidity."[58] But in an inversion, however, the death that hunts down European bodies was also a *capacity*, turned into resilience that enabled slaves to thrive in otherwise degenerative climates, as "men entirely exempt from the debilitating influence of the climate, men to whom the yellow fever is unknown . . . and far more agile by constitution and habits."[59] Resistant to the death that plagues Europeans, but identified climatologically and physiologically with the zones that bring it to their doorstep, black bodies become the geographic embodiment of necrotic intimacy. Like Humboldt's description of the sensitivity of indigenous peoples, where "'the character of the savage [*sauvage*] is modified everywhere by the nature of the climate and the soil where he lives,'" Stephen writes that "the soil itself is [the slave's] inexhaustible magazine; rapidly producing for him by the briefest and easiest culture, and even by its own spontaneous gift the esculent plants, and fruits, on which he well knows how to subsist."[60] Stephen's recommendations for a liberal policy of limited emancipation draws on this figurative quality of physiology, both colonial degeneration and racialized resilience. The global unity of empire and colonies is a physio-geography that brings "the abject cast" into the "social edifice" that includes degeneration and death within a progressive system.[61]

In lieu of a proper name, an essential figurability smudges together slave, savage, and native, configuring each through the vague and porous borders that turn fluidity into a fixed position within species classification. Stephen projects fluid bodies into the pre-civilizational past while making those bodies central to the future of the global security of empire. Slaves' vulnerability to, and their physiological intimacy with, geography and climate provide the grounds for racial classifications that, as Daniel Nemser has argued, operate by deciding "which bodies are fixed and which are fluid."[62]

Whereas other contemporary texts envisage freed slave colonies as a risk of contagion in which an unhealthy excitation would ruin the geopolitical balance of European powers, Stephen represents emancipation and incorporation as an almost inoculating protection against the risk of French invasion. While "the establishment of a negro state" in those colonies might be "fatal," far worse for the British empire is "that our sugar colonies should be . . . conquered by foreign arms."[63] In so doing, Stephen articulates a general idea of global power in which no part is rendered truly outside or exterior, producing a novel representation of liberal empire in relation to the abject cast and the deathly zones of the colonies. In this progressive, unfolding view of the globe, the task at hand is to assess which kinds of death are innate or foreign and which can be incorporated to preserve global power. Stephen's primary recommendation, then, is the incorporation of yet-to-be-freed slaves—those unnamed, indeterminate, and vulnerable bodies—and degenerative territories within the political order of Great Britain: "'Is it necessary then that large bodies of negro troops should be raised and maintained in Jamaica and other Islands?' If we would long retain the sovereignty over them; if we would prevent their soon swelling the dominions of the French Republic; that expedient, objectionable and hazardous though during our present situation of their brethren in those Islands it may be, I think it must be adopted."[64]

Such hospitable inclusion of degenerative zones and bodies, of the necrotic geography of the West Indies, preserves rather than transforms pre-existing colonial sovereignty. This part that had previously not been a part of the British empire now becomes "[t]he ground of *necessity*" that must be included to preserve the sovereign, global order. Indeed, the justification for this new, liberal policy relies on the physio-geography of the colonies as a means to the end of colonial governance, in which "the physical powers of negroes, opposed to those of Europeans in a hot climate" can be turned to "save the lives of our soldiers [and] to attain the end for which they have been hitherto sacrificed."[65] In order for this to be done, the conversion of slaves into something like English laborers, brought under the penal law of the British empire and remunerated through wage labor, must be achieved. As Stephen continues, "The foundation then, Sir, on which alone I deem it practicable to build on the future security of the sugar colonies is that of meliorating the condition of the great mass of the people, and converting them from dangerous enemies into defenders." Otherwise, "the slave system will continue to be a source of internal weakness and danger till revolution or foreign conquest become the well merited result."[66] Stephen concludes with a turn towards the new colonization of Trinidada, offering it as both "a farm of experiment, and a fortress [of] a new and happy system of colonization, which while it produces wealth, may with an equal progress furnish free, strong, and faithful hands to defend it."[67] This bright and healthy future is dependent upon the necropolitical inclusion of freedom for slaves *within* an absolute hierarchy of sovereignty. The West Indies, a geographic expression of black physiology, is the threshold that brings the outside inside.

The assemblage of soil and climate, disease and death confirm the historical indeterminacy of a site that not only can be included into empire but also can be grounds of its security and vitality. The general idea of sovereignty, for Coleridge expressed as Man and for Stephen as a new, progressive mode of empire, is stabilized by an indispensable indistinction between life and death, or a chiasmic dialectics of configuration and classification. What I am trying to suggest in this constellation of texts is that Romantic debates over life in 1816 were always-already contaminated by colonial figurations of the "*living dead*." This contamination is by no means particular to Coleridge, but extends also to the urtext of the vitalist debates, John Hunter's *A Treatise on the Blood, Inflammation, and Gun-shot Wounds*. There Hunter's vitalist principle is given as one in which the loss of any part cannot be considered a real loss; identity resides in a separable whole:

> a body may be so compounded, as to make a perfect whole of itself . . . Under those circumstances, the removing a portion is not taking away a constituent part, upon which the whole depends, or by which it is made into a whole, but is only taking away a portion of the whole; the remaining portion being in equal quality to the whole . . . because all its parts are similar, and in harmony with each other.[68]

Read beyond the individual body, such an account of unity makes possible the inclusion *and* exclusion of parts that extend to what Hunter, like Coleridge, calls "those inferior order of animals."[69] At once separable from the whole of life but constitutive to our understanding of its hierarchy, these forms are rendered vital through their expendability. The incorporation and distribution of the physical, dead, or mechanical within such a system *render forms nonessential.*

In the case of both Romantic vitalism and colonialism, it is perhaps more adequate to describe such operations not as inclusion and exclusion but rather as what Nemser calls a "*differential exteriorization* [of] [r]acialized bodies and social configurations."[70] Composed through such differential exteriorities, vulnerabilities, and intimacies with death, the inclusive or porous nature of Romantic classification appears less destabilizing or excessive and more as the stabilizing ground of hierarchical structure. Stephen's configuration of geography, history, and physiology attest to a necessary inclusion of what could be rendered as zones of living dead into the vital, global body politic of the British empire.

NOTES

Chapter title: Samuel Taylor Coleridge, "Hints Towards the Formation of a More Comprehensive Theory of Life," in *The Collected Works of Samuel Taylor Coleridge*, vol. 11, ed. H. J. Jackson and J. R. de J. Jackson, Bollingen Series (London and Princeton: Princeton University Press, 1995), 1030 (hereafter referred to as "Theory").

1. Achille Mbembe, "Necropolitics," trans. Libby Meintjes, *Public Culture* 15, no. 1 (2003): 11–40.
2. Alastair Hunt and Matthias Rudolf, "Introduction: The Romantic Rhetoric of Life," *Romanticism and Biopolitics, Romantic Circles* (December 2012), accessed March 31, 2019, https://romantic-circles.org/praxis/biopolitics/HTML/praxis.2012.hunt-rudolf.html.
3. Paul Youngquist, "Lyrical Bodies: Wordsworth's Physiological Aesthetics," *European Romantic Review* 10, no. 1–4 (1999): 152–162, accessed March 30, 2019, https://doi.org/10.1080/10509589908570073.
4. Catherine Gallagher, *The Body Economic: Life, Death, and Sensation in Political Economy and the Victorian Novel* (Princeton: Princeton University Press, 2006), 33.
5. Daniel Nemser, *Infrastructures of Race: Concentration and Biopolitics in Colonial Mexico*, (Austin: University of Texas Press/Border Hispanisms, 2017), 162.
6. Tilottama Rajan, "Organicism," *English Studies in Canada* 30, no. 4 (December 2004): 46–50, 47.
7. Jacques Khalip and Forest Pyle, "Introduction: The Present Darkness of Romanticism," in *Constellations of a Contemporary Romanticism*, ed. Jacques Khalip and Forest Pyle (New York: Fordham University Press, 2016).
8. On this point, see Alexander G. Weheliye, *Habeas Viscous: Racializing Assemblages, Biopolitics, and Black Feminist Theories of the Human*, (Durham: Duke University Press, 2014); see also Nemser, *Infrastructures of Race*, 162.
9. Frantz Fanon, *Wretched of the Earth*, trans. Richard Philcox (New York: Grove Press, 1961, rpt. 2004), 3 (my emphasis). In the English translation, the word "configuration" is used eight times and seems to refer both to a distinction and an articulation, to segregation and relationality. In certain contexts Fanon uses it to describe colonial dualism while in others it refers to the association of Black men with animals through phrenology or anatomy. While Fanon's emphasis throughout *Wretched* is clearly on the Manichean binaries that structure colonial space and life, the chain of associations it creates in order to stabilize "a world divided in two" evokes this more chiasmic relation I am trying to draw out in his and Mbembe's work.
10. Mbembe, "Necropolitics," 17 (my emphasis).
11. Coleridge, "Theory of Life," 526, 512; James Stephen, *The Crisis of the Sugar Colonies; or, An Enquiry into the Objects and Probable Effects of the French Expedition to the West Indies*, (London: J. Hatchard, 1802), 202.
12. As Katherine Coburn's introduction to "Theory" clarifies, this was an intermediary text for Coleridge, written under the sway of German *Naturphilosophie* after which Coleridge turned even further towards theology. Nonetheless, it highlights an important cultural and political crux in debates over vitalism through its incorporation of nonhuman organisms and even inorganic matter into human sovereignty and autonomy.
13. Coleridge, "Theory," 508, 512.
14. Jacob Risinger, "Coleridge, Politics, and the Theory of Life," *Studies in English Literature* 55, no. 3 (Summer 2015): 647–667, 659.
15. Coleridge, "Theory," 1030.
16. Rajan, "Dis-figuring Reproduction: Natural History, Community, and the 1790s Novel," *CR: The New Centennial Review* 2, no. 3 (Fall 2002): 211–252, 217.
17. Rajan, "Dis-figuring Reproduction," 211.

18. Coleridge, "Theory," 538–539.
19. Coleridge, "Theory," 539.
20. Coleridge, "Theory," 495.
21. Coleridge, "Theory," 495; Coleridge's incorporation of the non-vital residue or relic of life is quite different than William Lawrence's approach, which follows Bichat's, and in which life was a force that resisted or was oppositional to death. For Lawrence, the absolute demarcation between inorganic, inert matter and organized, living bodies was that the latter "*grow by an internal power*, and finally perish by that internal principle, or by the effect of life itself." William Lawrence, *An Introduction to Comparative Anatomy and Physiology* (London: J. Callow, 1816), 62.
22. Coleridge, "Theory," 510–511.
23. Andrzej Warminski, "Introduction: Allegories of Reference," in Paul de Man, *Aesthetic Ideology*, ed. Andrzej Warminski (Minnesota: University of Minnesota Press, 1996), 11.
24. Coleridge, "Theory," 551.
25. My argument here is informed by Robert Bernasconi's arguments about preformationism and epigenesis in Kant; see "Kant and Blumenbach's Polyps: A Neglected Chapter in the History of the Concept of Race," in *The German Invention of Race*, ed. Sara Eigen and Mark Larrimore (Albany: SUNY Press, 2006), 73–90.
26. Coleridge, "Theory," 486.
27. Ian Balfour, *The Rhetoric of Romantic Prophecy* (Stanford: Stanford University Press, 2002), 283.
28. Coleridge, "Theory," 516, note 2.
29. Coleridge, "Theory," 539 (my emphasis).
30. Coleridge, "Theory," 512 (my emphasis).
31. Coleridge, "Theory," 493–494.
32. On this point see Giorgio Agamben, *The Use of Bodies*, trans. Adam Kotsko (Stanford: Stanford University Press, 2015), 195–206.
33. Coleridge, "Theory," 492.
34. Coleridge, "Theory," 537, note 3.
35. Coleridge, "Theory," 537, note 3; *OED*, s.v. "exuviae." The *OED* traces this term to the Latin "*exuĕre*" meaning "to divest oneself of" as well as "garments stripped off, skins of animals, the spoils of an enemy."
36. Coleridge, "Theory," 537, 516.
37. Coleridge, "Theory," 508.
38. Beatrice Hanssen, *Walter Benjamin's Other History: Of Stones, Animals, Human Beings, and Angels* (Berkeley and Los Angeles: University of California Press, 1993), 60.
39. Coleridge, "Theory," 493.
40. Coleridge, "Theory," 511.
41. Walter Benjamin, *The Arcades Project*, trans. Howard Eiland and Kevin McLaughlin (Cambridge and London: Belknap Press of Harvard University Press, 1999), 474.
42. On the colonialist origins of natural history, see Mary Louise Pratt, *Imperial Eyes: Travel Writing and Transculturation* (London and New York: Routledge, 1992). For a consideration of natural history in Benjamin's work, which is highly complementary to recent work on colonialist natural history, see Hanssen, *Walter Benjamin's Other History*.
43. Gavin Walker, "Primitive Accumulation and the Formation of Difference: On Marx and Schmitt," *Rethinking Marxism* 23, no. 3 (2011): 384–404, 388.
44. Evan Gottlieb, *Romantic Globalism: British Literature and the Modern World Order, 1750–1830* (Columbus: Ohio State University Press, 2014), 10. My turn to global politics here is meant in part to consider how such horizontality was configured within colonialist apologetics, but also as a way to situate recent treatments of sympathy as a physiological term of

radical politics within a larger colonial context. In considering recent appeals to a politics-motivated sympathy, Peter DeGabriele's argument that eighteenth-century imperialist sovereignty was maintained through the spatialization of sympathy within the borders of the nation and direct violence and domination in colonial territories is instructive. See DeGabriele, "Sympathy for the Sovereign: Sovereignty, Sympathy, and the Colonial Relation in Edward Gibbon's 'The Decline and Fall of the Roman Empire,'" *The Eighteenth Century* 53, no. 1 (2012): 1–22.

45. See Weheliye, *Habeas Viscous* and Nemser, *Infrastructures of Race* on this point.
46. Stephen, *Crisis*, 202.
47. Slavery and colonial plantations were, of course, central to the British empire's economic primacy throughout the Romantic period. As Stephen and others understood, British finance capital would collapse without the plantation.
48. Thus, as he makes clear, *Crisis*, 27, one of Stephen's primary goals in publishing his letters is to "fi[x], the true nature of that condition inadequately defined by the term *slavery* from which the French negroes have passed, and its essential difference from that to which they have attained."
49. Rajan, "Dis-figuring," 211.
50. Stephen, *Crisis,* 26.
51. Stephen, *Crisis,* 27 (my emphasis).
52. Stephen, *Crisis,* 27.
53. Stephen, *Crisis*, 49. As Jordana Rosenberg has noted, this distinction between "African bond-laborers and European proletarians in the colonies [was used] to disrupt labor alliances and secure social order toward capital accumulation in the colonial Atlantic." *Critical Enthusiasm: Capital Accumulation and the Transformation of Religious Passion* (Oxford: Oxford University Press, 2011), 107.
54. Stephen, *Crisis*, 53.
55. Stephen, *Crisis*, 54.
56. Mbembe, "Necropolitics," 13.
57. Examples of this diagnosis exceed citation. But John Gabriel Stedman's *Narrative of a Five Years' Expedition; against the Revolted Negroes of Surinam in Guiana* (London: J. Johnson and J. Edwards, 1796) and Henry Brougham's *An Inquiry into the Colonial Policy of the European Powers* (Edinburgh: D. Willison, 1803) discuss it as a central obstacle to military victory in the West Indies.
58. Stephen, *Crisis*, 61.
59. Stephen, *Crisis*, 64. I take this concept of vulnerability from Nemser, *Infrastructures*, 133–164.
60. Alexander von Humboldt, quoted in Nemser, *Infrastructures*, 160; Stephen, *Crisis*, 64–65.
61. Stephen, *Crisis*, 120.
62. Nemser, *Infrastructures,* 164.
63. Stephen, *Crisis*, 79.
64. Stephen, *Crisis*, 118.
65. Stephen, *Crisis*, 120.
66. Stephen, *Crisis*, 150.
67. Stephen, *Crisis*, 202
68. John Hunter, *A Treatise on the Blood, Inflammation, and Gun-shot Wounds* (London: John Richardson, 1794), 84. *Eighteenth Century Collections Online*, Gale, CW108342200, accessed March 31, 2019.
69. Hunter, *Treatise*, 89.
70. Nemser, *Infrastructures*, 162 (my emphasis).

10

THE METAPHYSICAL MACHINERY OF MINING IN NOVALIS'S WORKS

Christina M. Weiler

Mining presents an important facet of the imaginative designs of Romanticism. While mining does not include automata in the prototypical sense, it shares with automated humanoids the ability to highlight a nonprivileged position of the human over the material. Underground in the mine, human bones fossilize together with plant and animal remains into inorganic stone, levelling perceived ontological hierarchies. As a liminal space, the mine challenges definitions of the human as detached from nature and provokes fears of a loss of self, as the definitions of what it means to be human are corroded. The dangers and difficulties of laboring underground further contest the idea of human superiority. Stories set under the earth highlight the fragility of the human both on an organic and on a conceptual level. The human does not dominate underground since human lives, and with them our definition of the human, are put at risk. In addition, mining draws attention to our embeddedness in the natural environment and our relations with it.

While mining today is associated with the use of big machinery, the automation of mining was still in its early stages during the Romantic period. Mining was at a juncture during this era. It was still deeply rooted in traditional practices while technological and scientific innovations were beginning to change the world. These innovations provoked concerns about the instrumentalization and objectification of nature. Stories about Romantic automata imagine the consequences of such developments, often anticipating a dystopian future. In contrast, Romantic images of mining allude to an idealized past as a positive counter model, especially during the early stages of German Romanticism.[1]

Friedrich von Hardenberg (1772–1801), alias Novalis, was not only one of the defining authors of Early German Romanticism but also an expert in mining. He studied mining engineering and geology at the Mining Academy of Freiberg from 1798–1799, and he worked as an assessor in a salt mine in Weißenfels, Saxony, which his father oversaw. Thus, it is fair to say that he was intricately familiar with the machinery of mining. In his writings Novalis portrays both the actual practice of mining, which represented an important economic sector during the Romantic period, and the idealized Romantic image of mining as an old, sacred craft that enables metaphysical insights. His professional work in the mining business runs chronologically parallel to his literary creations, but ideologically the two spheres stand in tension with each other. Romantic mining in Novalis's works negotiates questions about economic gain and creative aesthetics in a period of rapid mechanical innovations. Unlike for-profit mining, aesthetic mining in his novels functions as a metaphysical practice that synthesizes artisanal craft, religion, and natural philosophy and thus reflects the Early German Romantic aesthetic ideal of *Universalpoesie*.

Novalis's pragmatic notes from his work as a saline assessor show that his professional objective was to optimize the mining process from a financial perspective. His notes list material objects, technical numbers, and costs. In a protocol of an assessment of a saline mine from June 11, 1799, he writes, for example: "Bey Commissarischer Untersuchung des Oeconomieplans für die Saline Kösen fand sich, daß 42 000 Stück Salz . . . jährlich mit Vortheil gesotten und die dazu erforderliche Soole zu dem Grade von 3 ¾, gradirt, auch damit die Bedürfnisse der Coctur Niederlage bestritten werden können."[2] [During the provisional examination of the business plan for the saline mine Kösen, it was found that 42,000 units of salt . . . can be profitably extracted with a yearly net profit and that the brine necessary for this be graded to the degree of 3 ¾, thus also meeting the requirements of the processing facility.] Novalis's professional notes thus indicate that he was an expert in the technical intricacies of saline mining and that his position required him to assess mining as a business operation. His objective as an assessor was materialistic, not metaphysical.

The beginning of the industrialization led to a growing automatization and exploitation of labor.[3] Mining machines were on their way to becoming automated tools employed to deplete natural resources in the developing modern capitalist economy. Novalis did not experience the height of the industrialization in Europe during his lifetime.[4] It was not until the nineteenth century that the steam engine significantly increased the demand for mining products and simultaneously fostered the development of more efficient mining machinery.[5] He did not personally witness changes in the mining process on a larger scale wrought by the steam engine, but he learned about this new technology in Freiberg.[6] The looming transformations that the invention of the steam engine heralded might have fueled his longing for the artisanal mining craft of an idealized past.

Next to technological developments, Novalis also saw the beginnings of the industrial capitalization reflected in the mining sector. The profit-driven exploitation of laborers developed early on in for-profit mining, which was an important precondition for the industrialization.[7] In a letter of April 28, 1800, reporting on the business of coal mining to his teacher Abraham Werner, Novalis discusses the impact of the work on the health of the miners, noteworthy, in the context of profit calculations: "Die Seltenheit der Leute, und die Nothwendigkeit sie wieder zu haben, macht es nöthig, sie im Sommer so viel verdienen zu lassen, daß sie zum Theil noch den Winter durch davon leben können, wo wenig Verdienst ist . . . auch ist die Arbeit äuserst beschwerlich, schmuzig und ungesund. Hautschäden und Gichtübel sind unter diesen Leuten sehr häufig" (*Schriften* III, 787). [The scarcity of people and the necessity to have them again make it necessary to let them earn so much in the summer that they can partially still live on it through the winter, when there is little to earn . . . the work is also extremely onerous, dirty, and unhealthy. Skin diseases and gout maladies are very common among these people.] The concern for the workers expressed by Novalis in his letter is primarily economically motivated. This reflects that in the business of mining, in which Novalis was actively engaged, human lives were increasingly objectified and seen as parts of monetary calculations.

As a mining assessor, Novalis was focused on the goal of increasing financial gains, and he was aware that the scientific advances in geology were serving the same purpose. In this context, Novalis viewed for-profit mining critically: "Erwerbsbergbau—wissenschaftlicher, geognostischer Bergbau. Kann es auch einen schönen Bergbau geben?" (*Schriften* II, 543) [For-profit mining—scientific, geognostic mining. Can there also be beautiful mining?] His statement shows that, for him, profit-oriented mining included geognostic mining, the scientifically oriented mining of geologists. Mining for business purposes was not concerned with beauty and thus unsatisfactory from the perspective of Romantic aesthetics.[8] Nevertheless and somewhat paradoxically, Novalis also saw an aesthetic potential in an alternative, artisanal mining practice rooted in an idealized past.[9] Novalis imagines artisanal mining as an intimate integration of the miner into nature, revealing the earth's beautiful treasures through his craftsmanship. In contrast to Novalis's prosaic notes from his time as a saline assessor, his literary works romanticize the mining practice. His novels explore the aesthetic potential of mining that for-profit mining neglects. Instead of analytic lists of numbers and materials, they present an exploration of conceptual connections and metaphorical synthesis. In his literary works, Novalis envisions artisanal mining as a metaphysical tool that catalyzes epistemological and phenomenological insights.[10] For Novalis, the poetic mine is a transgressive, liminal space.[11] It allows the human being to go beyond the surface of the natural world and of the self, and it enables an aesthetic perception of natural history as a harmonious cycle of transformation. The harmony of natural history is reflected in the mineral world and becomes thus visible underground, inscribed in the stones.[12]

Novalis's novel fragment *Heinrich von Ofterdingen*, first published in 1802, explores the metaphysical potential of mining as a "seltne, geheimnißvolle Kunst" (*Schriften* I, 241) [rare, mysterious art (*Henry*, 65)]. The *Bildungsroman* focuses on the life journey of its eponymous hero Heinrich von Ofterdingen. It presents a notion of mining as an art that points back to an understanding of mining as a craft in an idealized version of the Middle Ages, in which the novel is set. In the fifth chapter, Heinrich and his travel companions encounter an old miner on their journey. The miner is introduced as "ei[n] alte[r] Mann . . . , der . . . freundlich die neugierigen Fragen beantwortete, die an ihn geschahen" (*Schriften* I, 239) [an old man . . . who . . . amiably answered the questions they eagerly put to him (*Henry*, 64)]. This description introduces the miner as a man of wisdom. It is not bodily strength that is highlighted in his portrayal but the experience of old age and the knowledge to be able to answer questions. The emphasis is not on physical labor or material wealth but on metaphysical insight. The riches that the "Schatzgräber" (*Schriften* I, 239) [treasure-grubber (*Henry*, 64)] unearths are not only of material but primarily of metaphysical value. The old miner has a mysterious aura, radiating knowledge and power: "Er sprach aber sehr bescheiden von seinen Kenntnissen und seiner Macht, doch trugen seine Erzählungen das Gepräge der Seltsamheit und Neuheit" (*Schriften* I, 239). [But he spoke very modestly of his knowledge and power; still his talk bore the stamp of strangeness and novelty (*Henry*, 64).] The metaphysical insights that the miner has gained through his work have set him apart from other people.

The metaphysical value of mining is reflected in a strong conceptual connection between mining and religion in Novalis's *Ofterdingen*. Echoing medieval beliefs, mining is presented as a sacred practice.[13] The old miner was driven by a longing for metaphysical insights about the numinous in nature when he set out to join the mining profession. He explains that he became interested in mining not out of economic ambition but out of religious devotion. He was intrinsically motivated to explore the mineral world through his adoration of precious metals and jewels on religious artifacts and his curiosity about their origin:[14]

> Von Jugend auf habe er eine heftige Neugierde gehabt zu wissen, was in den Bergen verborgen seyn müsse, wo das Wasser in den Quellen herkomme, und wo das Gold und Silber und die köstlichen Steine gefunden würden, die den Menschen so unwiderstehlich an sich zögen. Er habe in der nahen Klosterkirche oft diese festen Lichter an den Bildern und Reliquien betrachtet, und nur gewünscht, daß sie zu ihm reden

> könnten, um ihm von ihrer geheimnißvollen Herkunft zu erzählen. (*Schriften* I, 239–40)
>
> [From youth on he had had a great curiosity to know what might be buried in the mountains, where the water in springs came from, and where the gold and silver and precious stone were found which so irresistibly attract human beings. He had often contemplated these sparkling solids on the icons and relics of the near-by monastery church and had wished that they could talk to him and tell him about their mysterious origin. (*Henry*, 64)]

The description of the jewels on the religious artifacts as "feste Lichter" [sparkling solids] presents a striking light metaphor that gives a creative take on the conventional metaphor of light as a representation of insight. The "feste Lichter" evoke the notion of insights materialized in stone. Through the religious context, the insights that the material jewels hold for the miner gain a metaphysical dimension. This is emphasized through the miner's wish to have the jewels speak to him so that he can uncover their divine secrets.

The conceptual link between mining and the divine is further strengthened when the old miner describes seeing a mine for the first time as a metaphysical experience: "mit unglaublicher Neugierde und voll stiller Andacht stand ich bald auf einem solchen Haufen, den man Halde nennt, vor den dunklen Tiefen, die im Innern der Hütten steil in den Berg hineinführten" (*Schriften* I, 240) [with unbelievable curiosity and full of quiet reverence I soon stood on one of those piles, which they call 'barrows,' in front of the dark shafts which led from the inside of the huts steeply down into the mountain (*Henry*, 65)]. The miner emphasizes his curiosity and reverence in the liminal space of the mine's entrance. Before descending into a mine for the first time, he remembers attending a mass read by a monk asking for divine protection of the miners. During this mass, he had a stronger religious experience than ever before in his life. "Ich hatte nie mit mehr Inbrunst gebetet, und nie die hohe Bedeutung der Messe lebhafter empfunden" (*Schriften* I, 241). [I had never prayed more fervently and never experienced the high significance of the mass more keenly (*Henry*, 66).] The physical space of the mine intensifies religious worship. The old man remembers vividly the adoration with which he perceived the other miners in this context. He saw them as heroic seekers of knowledge.

> Meine künftigen Genossen kamen mir wie unterirdische Helden vor, die tausend Gefahren zu überwinden hätten, aber auch ein beneidenswerthes Glück an ihren wunderbaren Kenntnissen besäßen, und in dem ernsten, stillen Umgange mit den uralten Felsensöhnen der Natur, in ihren

> dunkeln wunderbaren Kammern, zum Empfängniß himmlischer Gaben und zur freudigen Erhebung über die Welt und ihre Bedrängnisse ausgerüstet würden. (*Schriften* I, 241)
>
> [My future companions appeared to me like subterranean heroes who had to overcome a thousand dangers but also possessed an enviably happy lot in their wonderful knowledge. And in their quiet, earnest association with the primeval rocks within nature's dark and marvelous chambers it seemed they were equipped to receive heavenly gifts and to be joyfully exalted above the earth with its afflictions. (*Henry*, 66)]

The miner recalls the great respect he felt for his future fellow craftsmen due to their metaphysical connection to stones as representatives of the divine in nature. Although the miners physically pursue their work underground, they are metaphysically elevated through their deeper philosophical understanding of the cosmos, which their profession enables.

The tools used for mining emphasize its metaphysical dimension in the novel. As a mining apprentice, the miner received a lamp and a wooden cross as his first tools. "Der Steiger gab mir nach geendigtem Gottesdienst eine Lampe und ein kleines hölzernes Krucifix, und ging mit mir nach dem Schachte, wie wir die schroffen Eingänge in die unterirdischen Gebäude zu nennen pflegen" (*Schriften* I, 241). [At the end of the devotion, the foreman gave me a lamp and a small wooden crucifix and went with me to the shaft, as we are wont to call the steep entrances to the subterranean structures (*Henry*, 66).] The miners' tools are philosophical instruments as they aid in the metaphysical experience of nature.[15] What enables the hermeneutic process of deciphering nature are the miners' tools: the cross, as a reminder of the divine origin of nature, and the lamp, which illuminates the darkness underground and thus makes insight possible. It is noteworthy that the lamp metonymically alludes to visionary perception while the cross, as distinctly wooden, metonymically represents the organic. The emphasis on light and the organic in the description of the philosophical mining instruments evokes the metaphysical revelations uncovered by the aesthetic mining organon about the interconnectedness of the organic and the mineral sphere through the numinous. The focus in the representation of mining in the novel is not on innovative technology but on tradition and the connection between human and nature. The goal of the aesthetic mining practice is not to exploit but to explore nature.

Mining in Novalis's *Ofterdingen* is described as an explorative endeavor that connects the miner with the cosmos. The old miner describes how he was initiated into the sphere of the mine by the pit foreman in spiritual terms:

> Er fuhr voraus, und schurrte auf dem runden Balken hinunter, indem er sich mit der einen Hand an einem Seil anhielt das in einem Knoten an einer Seitenstange fortglitschte, und mit der andern die brennende Lampe trug. Mir war seltsam feyerlich zu Muthe, und das vordere Licht funkelte wie ein glücklicher Stern, der mir den Weg zu den verborgenen Schatzkammern der Natur zeigte. (*Schriften* I, 242)
>
> [He went ahead and glided down the round beam, carrying the lit lamp in one hand and holding with the other to a rope which slid down a side pole. . . . I had a peculiarly solemn feeling, and the light ahead twinkled like a lucky star showing me the way to the hidden treasure chambers of nature. (*Henry*, 66)]

The image of the joyful guiding star has religious connotations as it evokes the star guiding the three kings to the Child Jesus. Furthermore, the image connects distant poles of nature as it links the lamp underground with the stars in the sky. The image gives the vertical movement of the miner downwards into the mine a cosmological dimension, connecting the inner earth metaphorically with its upward antipode, the sky.

Aesthetic mining, as it is presented in the novel, is not merely a bread-winning profession but a calling. The old miner explains to Heinrich that his wish to become a miner was motivated by an innate desire: "Es läßt sich auch diese volle Befriedigung eines angebornen Wunsches, diese wundersame Freude an Dingen, die ein näheres Verhältniß zu unserm geheimen Daseyn haben mögen, zu Beschäftigungen, für die man von der Wiege an bestimmt und ausgerüstet ist, nicht erklären und beschreiben" (*Schriften* I, 242). [Nor is it possible to describe or explain this complete satisfaction of an inborn desire, this wondrous joy in things, which may have a close relation to our mysterious existence, to activities one is prepared and predestined for from the cradle (*Henry*, 67).] The miner believes that he has been destined from birth for his work underground. The notion of mining as a calling is rooted in the metaphysical dimension of the miners' work. Unlike for-profit mining, their artisanal mining practice seeks insight into nature and through it into the divine. The joy that the miners feel when they follow their calling results from their recognition of divine wisdom and providence being present in nature: "der Bergbau muß von Gott gesegnet werden! denn es giebt keine Kunst, die ihre Theilhaber glücklicher und edler machte, die mehr den Glauben an eine himmlische Weisheit und Fügung erweckte, und die Unschuld und Kindlichkeit des Herzens reiner erhielte, als der Bergbau" (*Schriften* I, 244) [mining must bear God's blessing, for there is no art which might make its participants happier and nobler, which would do more to arouse men's faith in a heavenly wisdom and providence, and which would keep the innocence and childlikeness of the heart in greater purity, than mining (*Henry*, 69)]. Mining preserves innocence and has a purifying effect because it brings man in contact with the divine.

The miners value the riches under the earth not for their financial worth but for the philosophical insights they allow about history and nature:

> Arm wird der Bergmann geboren, und arm gehet er wieder dahin. Er begnügt sich zu wissen, wo die metallischen Mächte gefunden werden, und sie zu Tage zu fördern; aber ihr blendender Glanz vermag nichts über sein lautres Herz. Unentzündet von gefährlichem Wahnsinn, freut er sich mehr über ihre wunderlichen Bildungen und die Seltsamkeiten ihrer Herkunft und ihrer Wohnungen, als über ihren alles verheißenden Besitz. Sie haben für ihn keinen Reiz mehr, wenn sie Waaren geworden sind, und er sucht sie lieber unter tausend Gefahren und Mühseligkeiten in den Vesten der Erde, als daß er ihrem Rufe in die Welt folgen, und auf der Oberfläche des Bodens durch täuschende, hinterlistige Künste nach ihnen trachten sollte. (*Schriften* I, 244–245)

> [The miner is born poor and he dies poor. He is content to know where the metal powers are found and to bring them to the light of day, but their dazzling glamor has no power over his pure heart. Uninflamed by perilous frenzy, he takes more delight in their peculiar structures and their strange origin and habitat than in their possession which promises so much. They have no charm for him any more once they are turned into commercial articles, and he had rather look for them within the strongholds of the earth amid a thousand dangers and drudgeries than to follow their call into the world and to strive after them up on the surface by means of deluding, deceitful arts. (*Henry*, 70)]

The old miner explains that he is not driven by financial greed but following his calling.[16] The insights about nature and the self, made possible through mining, are an answer to his curiosity, which in turn leads to a profound contentment that cannot be reached through financial gains. This demonstrates a critique of the capitalization of human labor and natural resources. If mining is reduced to a merely utilitarian business transaction, the human miner and the mineral sphere become mere means to an end. The minerals and metals are robbed of their connection to the divine when they are extracted; they are taken from the earth as tradable products and thus become alienated from their place in natural harmony.

Mining in Novalis's *Ofterdingen* is a tool for understanding human existence. The old miner reveals to Heinrich that mining is a hidden symbol of human life: "Wahrhaftig, das muß ein göttlicher Mann gewesen seyn, der den Menschen zuerst die edle Kunst das Bergbaus gelehrt, und in dem Schooße der Felsen dieses ernste Sinnbild des menschlichen Lebens verborgen hat" (*Schriften* I, 246). [Indeed he must have been a godly man who first taught men the noble art of mining and concealed this earnest symbol of man's life in the bosom of the mountains (*Henry*, 71).] The labyrinth of tunnels is, according to the old miner, an analogy to the possible paths of life and the unpredictable nature of the stones reflects the element of coincidence in human existence. The miner explains that mining invites a close examination of the earth, both in the sense of exploring it underground and of traveling across the land. Thus, it is both a vertical and a horizontal journey in which spatial movement increases insight. "Unsere Kunst macht es fast nöthig, daß man sich weit auf dem Erdboden umsieht. . . . Sie hat überall mit andern Schwierigkeiten zu kämpfen gehabt, und da immer das Bedürfniß den menschlichen Geist zu klugen Erfindungen gereitzt" (*Schriften* I, 260). [Our work almost requires one to look around far and wide on the earth. . . . It has had to struggle with special difficulties everywhere; and since necessity has always been the mother of clever inventions (*Henry*, 86).] The manifold challenges of mining, which represent life challenges more generally, stimulate human abilities and experiences. Mining can thus be seen as a catalyst for the progressive advancement of humanity.

The miner leads Heinrich and his travel companions into a system of tunnels evoking the metaphorical labyrinth of his analogy. In these caves, they encounter bones of ancient animals, which bring them into material contact with natural history: "was die Aufmerksamkeit Aller vorzüglich beschäftigte, war die unzählige Menge von Knochen und Zähnen, die den Boden bedeckten. Viele waren völlig erhalten, an andern sah man Spuren der Verwesung, und die, welche aus den Wänden hin und wieder hervorragten, schienen steinartig geworden zu seyn" (*Schriften* I, 253). [what engaged the attention of all particularly was the countless number of bones and teeth which covered the floor. Many were fully preserved; others had marks of disintegration, and those that stuck out of the wall here and there appeared to be petrified (*Henry*, 78).] The fossils are the physical representation of the interconnectedness of the organic and inorganic world. Human, animal, and plant remains become stone through fossilization whereby the organic physically changes into the inorganic.[17] This recognition establishes an innate physical bond between humans and the mineral sphere as different but permeable manifestations of nature.

Mining transgresses the boundaries between the organic and the inorganic. Even the mineralogical treasures, which the miners are unearthing, seem to grow plantlike out of the ground.[18] The old miner compares the beauty of the metals and gem stones he encounters to a magical garden:

> An manchen Orten sah ich mich, wie in einem Zaubergarten. Was ich ansah, war von köstlichen Metallen und auf das kunstreichste gebildet. In den zierlichen Locken und Ästen des Silbers hingen glänzende, rubinrothe, durchsichtige Früchte, und die schweren Bäumchen standen

> auf krystallenem Grunde, der ganz unnachahmlich ausgearbeitet war. (*Schriften* I, 262)
>
> [In some places I found myself in a veritable garden of magic. What I beheld was formed most artistically out of the most precious metals. In the elegant braids and branches of silver there hung sparkling, ruby-red, transparent fruits, and the heavy trees were standing on a crystal base inimitably wrought. (*Henry*, 88)]

The garden of gem stones and precious metals is a beautiful image that follows the Romantic aesthetic ideal of synthesizing harmony.[19] The organic and the mineral become permeable in the sphere of the mine in Novalis's novel.[20]

The epistemological insights about natural history that the subterranean sphere offers are complemented by the phenomenological recognition that humankind is part of the earth. Thus, the work of investigating the earth underground, which is inherent to mining, can be seen as a method of natural philosophy that creates awareness for the harmonious aesthetics of nature and its interconnectedness with mankind.[21] For Heinrich, the journey into the earth allows him to see himself as part of this organic harmony. He gains insights into the hidden mechanics of nature and into his own life as a predetermined part in the universe's greater cyclical mechanisms. In one of the caves, Heinrich and his companions encounter a hermit, who constitutes a parallel figure to the miner. Like the miner, the hermit has decided to spend his life underground because this space facilitates reflections about life. The hermit uses the metaphorical image of reversed astrologers to describe the miners' practice of natural history and philosophy underground:

> Ihr seyd beynah verkehrte Astrologen. . . . Wenn diese den Himmel unverwandt betrachten und seine unermeßlichen Räume durchirren: so wendet ihr euren Blick auf den Erdboden, und erforscht seinen Bau. Jene studieren die Kräfte und Einflüsse der Gestirne, und ihr untersucht die Kräfte der Felsen und Berge, und die mannichfaltigen Wirkungen der Erd- und Steinschichten. Jenen ist der Himmel das Buch der Zukunft, während euch die Erde Denkmale der Urwelt zeigt. (*Schriften* I, 260)
>
> [You miners are almost astrologers in reverse. . . . Whereas they gaze incessantly at the heavens and stray through those immeasurable spaces, you turn your gaze into the earth and explore its structure. They study the powers and influence of the constellations, and you investigate the powers of rocks and mountains and the manifold effects of the strata of earth and rock. To them the sky is the book of the future, while to you the earth reveals monuments of the primeval world. (*Henry*, 86)]

The image of the reversed astrologer metaphorically depicts the metaphysical practice of artisanal mining as it is presented in Novalis's *Ofterdingen*. While astrologers look into the sky to predict the future, miners look underground where they can see the past. The vertical antipodes, sky and underground, reveal the mysteries of the universe.

In the hermit's cave, Heinrich finds a mysterious book written in a foreign language. This book contains images depicting scenes from his own life that simultaneously represent Heinrich's past and present and predict his future. "Es hatte keinen Titel, doch fand er noch beym Suchen einige Bilder. Sie dünkten ihm ganz wunderbar bekannt, und wie er recht zusah entdeckte er seine eigene Gestalt ziemlich kenntlich unter den Figuren" (*Schriften* I, 264). [It had no title, but as he looked through it, he found several pictures. They seemed wonderfully familiar to him, and as he looked more sharply, he discovered a rather clear picture of himself among the figures (*Henry*, 90).] The images that Heinrich encounters in the cave aid him on his journey to phenomenological insight because they allow him to synthesize his past, present, and future in a cosmological unity that implies divine providence. The images in the book mirror the cyphers of the fossils and metals embedded in the mountain insofar as both reveal a complex connection of life across time.

The subterranean sphere also serves as a catalyst for a prophetic vision, Heinrich's famous dream of the blue flower at the beginning of the novel. Before Heinrich sets out on his journey to discover the world and himself, he has a dream that anticipates his descent into the cave with the miner. He dreams of entering a blue cave: "Die Wände der Höhle waren mit dieser Flüssigkeit überzogen, die nicht heiß, sondern kühl war, und an den Wänden nur ein mattes, bläuliches Licht von sich warf" (*Schriften* I, 196). [This fluid, which was not hot but cool, covered the walls of the cave, where it emitted only a faint bluish light (*Henry*, 16).] In this dream cave, Heinrich has a sexual vision of the water underground materializing into young women while he is swimming in it. It is this experience in the cave that triggers his vision of the blue flower. The blue flower is a dream image embedded in a dream in the dream of the blue cave, twice removed from quotidian perception. The liminal space of the cave functions as a gate to metaphysical insight. Heinrich's dream of the blue flower, a hybrid being which is both plant and human, allows him to transgress the boundaries of natural categories and thus reach a more profound, harmonious connection with nature.[22] The dream of the blue flower also anticipates Heinrich falling in love with Mathilde. The metaphorical connection between his human beloved, Mathilde, the blue flower, and the stone cave, as a representation of the mineral world, strengthens the idea of the permeability of the categorization of nature.[23]

Novalis's *Ofterdingen* associates the craft of mining with the creative visionary work of the poet, for which Heinrich is destined. What the poet and the miner have in common is an extraordinary perception for the world that is reflected in their elevated subjective appreciation for its beauty. The merchants, who travel with Heinrich, metaphorically define poesy as a form of metaphysical mining:

> so erfüllt der Dichter das inwendige Heiligthum des Gemüths mit neuen, wunderbaren und gefälligen Gedanken. . . . Wie aus tiefen Höhlen steigen alte und künftige Zeiten, unzählige Menschen, *wunderbare* Gegenden, und die seltsamsten Begebenheiten in uns herauf, und entreißen uns der bekannten Gegenwart. (*Schriften* I, 210; italics in original)
>
> [so the poet fills the inner sanctuary of the spirit with new, wonderful and pleasing thoughts. . . . Within us as out of deep caverns there rise ancient and future times, countless, people, *marvelous* regions, and the strangest occurrences, snatching us away from the familiar present. (*Henry*, 31–32)]

The poet is metaphorically conceptualized as a miner who lifts poetic treasures out of the deep caves of his mind. He has a special talent to delight others with his beautiful creative thoughts. The poet allows us to see what would otherwise be hidden as does the miner, who brings to light the hidden beauty of the earth.

Both miner and poet are able to read nature's "große Chiffernschrift" (*Schriften* I, 79) [great cypher (*Novices*, 3)] which Novalis describes in his novel fragment *Die Lehrlinge zu Sais* (1802). These cyphers are inscribed, among others, "in Krystallen und in Steinbildungen, . . . im Innern und Äußern der Gebirge" (*Schriften*, I, 79) [in crystals and in stone formations, . . . on the inside and outside of mountains (*Novices*, 3)]. Thus, the mineral world presents one access point to the mysteries of nature and of the self.[24] Hyacinth, the protagonist of the fairy tale embedded in *Die Lehrlinge zu Sais*, sets out on his path to discover himself when he descends into mine shafts guided by a strange old sorcerer, a teacher figure that evokes the association of the old miner in Novalis's *Ofterdingen*.

> So viel man nachher vernahm, so hat er viel von fremden Ländern, unbekannten Gegenden, von erstaunlich wunderbaren Sachen erzählt, und ist drey Tage dageblieben, und mit Hyacinth in tiefe Schachten hinuntergekrochen. Rosenblüthchen hat genug den alten Hexenmeister verwünscht. . . . Endlich hat jener sich fortgemacht, doch dem Hyacinth ein Büchelchen dagelassen, das kein Mensch lesen konnte. (*Schriften* I, 93)
>
> [As far as could later be learned, the man told many stories of strange lands, of unknown regions, of marvelous things; he stayed for three days, and he and Hyacinth went down into deep caverns. Rose Petal cursed the old sorcerer. . . . At last the man went on his way, but he left Hyacinth a little book which no one could read. (*Novices*, 59–61)]

The illegible book that Hyacinth is presented with after his descent underground mirrors the book that Heinrich encounters in the cave. Nature's cyphers under the earth serve as a catalyst that promotes metaphysical insights about the self and nature, represented through Hyacinth's and Heinrich's journeys in Novalis's two unfinished novels.

Novalis's works negotiate questions about aesthetics and economics through mining. His manifold texts represent both metaphysical and material aspects of mining. The metaphysical machinery of mining in his literary work contrasts with the pragmatic mining assessor notes. Out of the materialism of for-profit mining, he developed an alternative aesthetic idea of mining in harmony with Romantic ideals. In Novalis's novels, mining is an epistemological and phenomenological practice that allows the one descending into the earth to follow his curiosity and to enter into deeper knowledge. Entering into the earth is a metaphor in Novalis's work for a process of deeper recognition of the self and world. Going underground allows the human being to come into closer contact with nature, which in turn allows insights about the world and the place of the human being in it. Knowledge about the earth enables phenomenological and epistemological insights. Thus, mining is for Novalis not only materialistic work, as he experienced it as a mining engineer, but also a metaphysical practice.

NOTES

Note on translations: For renderings into English of *Heinrich von Ofterdingen* (*Schriften* I, 183–334) and *Die Lehrlinge zu Saïs* (*Schriften* I, 71–110), I have used Palmer Hilty's and Ralph Manheim's translations respectively, referring to them by the short titles *Henry* and *Novices*; see Novalis, *Henry von Ofterdingen*, trans. Palmer Hilty (Prospect Heights: Waveland Press, 1990) and Novalis, *The Novices of Sais*, trans. Ralph Manheim, with illustrations by Paul Klee (Brooklyn, NY: Archipelago Books, 2005). All other translations in this chapter, unless otherwise indicated, are mine.

1. Helmut Gold points out the contrasting ways in which Romantic mining and automata motifs respond to modern societal developments: "Beide Motive reagieren auf eine moderne Erfahrung der Romantiker: Die Erfahrung von der Durchdringung aller Lebensbereiche mit dem Prinzip der Zweckrationalität, von Entfremdung und Verdinglichung. Der romantisch-literarische Einspruch beschritt dabei vor allem zwei Wege: den einer kritischen Antizipation und den eines konstruktiven Rückgriffs. Zum ersten gehören vor allem die vielfach untersuchten Automate-Motive, zum zweiten—neben anderen—die Bergbaumotive." [Both motives react to a modern experience of the Romantics: the experience of the permeation of all spheres of life with the principle of instrumental rationality, of alienation and objectification. Thereby, the Romantic-literary protest took above all two paths: one of critical anticipation and one of constructive recourse. The first one includes primarily the often examined automaton motifs, the second—among others—the mining motifs.] See Helmut Gold, *Erkenntnisse unter Tage: Bergbaumotive in der Literatur der Romantik* (Opladen: Westdeutscher Verlag, 1990), 7.
2. Novalis, *Schriften: Die Werke Friedrich von Hardenbergs,* ed. Richard Samuel et al., vol. 3 (Stuttgart: Kohlhammer, 1960), 743 (=*Schriften*, followed by the volume in Roman numerals and the page number in Arabic numbers).

3. Lieselotte Sauer highlights the connection between industrialization and automation. She states that "[g]rowing industrialization made the danger of man's degradation and his replacement by machines or automata even more apparent." See Lieselotte Sauer, "Romantic Automata," in *European Romanticism: Literary Cross-currents, Modes, and Models*, ed. Gerhart Hoffmeister (Detroit: Wayne State University Press, 1990), 291. While a large-scale replacement of miners through machines was not an imminent danger in Novalis's times, he was aware of the workers' degradation through modern mining practices.
4. Minsoo Kang points out that the Romantic period is chronologically situated before the height of industrialization. Kang notes that "[t]he technological revolution commenced in Britain in the second half of the eighteenth century. . . . But on the Continent, large-scale industrialization did not begin until decades later." See Minsoo Kang, *Sublime Dreams of Living Machines: The Automaton in the European Imagination* (Cambridge and London: Harvard University Press, 2011), 199.
5. See Gold, *Erkenntnisse unter Tage*, 43–44.
6. As Herbert Uerlings notes, "Berechnung und Konstruktion der damals modernen Dampfmaschinen . . . gehörten zum Freiberger Lehrstoff" [calculation and construction of the then modern steam engines . . . were part of Freiberg's curriculum (my translation)]. See Herbert Uerlings, "Novalis in Freiberg: Die Romantisierung des Bergbaus—Mit einem Blick auf Tiecks *Runenberg* und E.T.A. Hoffmann's *Bergwerke zu Falun*," *Aurora* 56 (1996): 57–77, 64.
7. Heather I. Sullivan situates "the dawn of the Anthropocene when mining's expanding influence aided by improved metallurgy provided greater access to (and use of) precious metals and, especially, coal, and thus allowed a literal re-shaping of the world by fueling the Industrial Revolution." See Heather I. Sullivan, "Dirty Nature: Ecocriticism and Tales of Extraction—Mining and Solar Power—in Goethe, Hoffmann, Verne, and Eschbach," *Colloquia Germanica* 44, no. 2 (2011): 111–131, 115.
8. Aesthetics are at the core of the Early Romantic project. Stefani Engelstein explains that "[e]merging from the intersection between art, natural history, and philosophy, the new field of aesthetics focused on the perception of beauty. . . . The word *aesthetics* was coined by Alexander Gottlieb Baumgarten, who initiated a philosophical debate with the publication of the *Aesthetica* in 1750." See Stefani Engelstein, *Anxious Anatomy: The Conception of the Human Form in Literary and Naturalist Discourse,* SUNY series, Studies in the Long Nineteenth Century (Albany: State University of New York Press, 2008), 6; see also the contribution by Engelstein in this volume.
9. As Matt Erlin puts it succinctly, the "representation of mining in German romantic literature can be read as an allegory of romantic aesthetics, and nowhere more so than in the work of Novalis." See Matt Erlin, *Necessary Luxuries: Books, Literature, and the Culture of Consumption in Germany, 1770–1815*, Signale: Modern German Letters, Cultures, and Thought, ed. Peter Uwe Hohendahl (Ithaca: Cornell University Press, 2014), 175.
10. As Theodore Ziolkowski argues, "[i]f, as Novalis . . . believed, stones, metals, and rock strata amount to transcriptions of the earth's history, what better place to study that history than in the mines and caverns of the earth, where the entire record is preserved and exposed?" See Theodore Ziolkowski, *German Romanticism and Its Institutions* (Princeton: Princeton University Press, 1990), 34.
11. Referring to E.T.A. Hoffmann's tale "Die Bergwerke zu Falun" ["The Mines of Falun"], Jens Klenner points out the liminality and transformative power of the mine in regard to the perception of nature and self: "The earth's surface becomes a liminal barrier or an inverting mirror and the miners are transformed when they break it open and pass through it." See Jens Klenner, "Geomorphic Poetics: Mountainous Transformations from E.T.A. Hoffmann to Elfriede Jelinek" (PhD diss., Princeton University, 2014), 72.

12. Dennis Mahoney stresses the importance of Novalis's professional mining work for the conception of *Heinrich von Ofterdingen* and highlights the harmonizing power of aesthetic mining in the novel: "die Arbeit am *Ofterdingen* fand . . . zu einer Zeit intensivster beruflicher Tätigkeit statt. . . . Wenn Novalis im *Ofterdingen* den Bergbau so sehr idealisierte, geschah das nicht aus einer Art Wirklichkeitsblindheit, sondern aus dem Wunsch . . . ein Modell für eine bessere, unentfremdete Welt zu liefern . . . . So wird für Novalis der Bergbau zum Sinnbild der romantischen Kunst, indem das Verschmelzen vom Alten und Neuen sowie das Entdecken und Hervorbringen des bisher Dunklen und Unzugänglichen eine wahrhafte Aufklärung und Versöhnung zwischen Menschen und Natur zustande bringen sollen" [the work on *Ofterdingen* took place . . . during a time of intensive professional activity. . . . When Novalis idealized mining so much in *Ofterdingen*, this occurred not due to a blindness to reality, but out of the wish . . . to provide a model for a better, nonalienated world. . . . Thus, mining becomes for Novalis the symbol of Romantic art, whereby merging of old and new as well as the discovery and creation of the hitherto obscure and inaccessible should achieve a true elucidation and reconciliation of human beings and nature]. See Dennis Mahoney, *Friedrich von Hardenberg (Novalis),* Sammlung Metzler 336 (Stuttgart: Metzler, 2001), 127–128.
13. As Hartmut Böhme states, "[d]er Bergbau selbst hatte im Mittelalter . . . einen durchaus sakralen Charakter. Überwiegend herrschte eine theozentrische Interpretation von Montanwissen und -technik. Die Erze, Mineralien, Reichtümer der inneren Erde dokumentierten die *magnalia dei*. Bergbau erhält so gottesdienstliche Bedeutung. . . . Es sind vor allem die Bergleute selbst, die eine Sakralkultur des Montanbaus entwickeln, deren Spuren noch bei Novalis . . . u.a. zu beobachten sind" [mining itself certainly had a sacred character in the Middle Ages. A theocentric interpretation of mining knowledge and technology predominantly prevailed. The ores, minerals, and riches of the inner earth documented the *magnalia dei*. Therefore, mining gains the significance of divine worship. . . . It is above all the miners themselves, who develop a sacral culture of mining, traces of which can still be observed in Novalis . . . and others]. See Hartmut Böhme, *Natur und Subjekt* (Frankfurt am Main: Suhrkamp, 1988), 68–69.
14. Kenneth S. Calhoon explains that that the miner's curiosity about the source of the precious metals and stones represents the philosophical search for origin more broadly. He states that the miner's "boyhood interest in mineralogy is characterized as a search for hidden origins . . . Novalis's novel thus makes the earth a *tertium comparationis* of manifold relevance, and geology appears as a sibling to those humanistic disciplines that seek to uncover the archaic strata of consciousness that have disappeared from view." See Kenneth S. Calhoon, *Fatherland: Novalis, Freud, and the Discipline of Romance*, Kritik: German Literary Theory and Cultural Studies Series (Detroit: Wayne State University Press, 1992), 107.
15. Thomas L. Hankins and Robert J. Silverman define the nature of philosophical instruments in the context of the study of nature: "But how should one consider philosophical instruments? Are they more like words, or more like things, or are they halfway in between? They are certainly things, but things whose purpose it is to help us analyze and reason about other things. They are things that we construct to represent and interpret nature." See Thomas L. Hankins and Robert J. Silverman, *Instruments and the Imagination* (Princeton: Princeton University Press, 1995), 9.
16. Erlin points out that the miner's warning against greed and commodification cannot be reduced to a modern critique of capitalism as it also recurs to earlier beliefs: "The real challenge presented by the passage, and by the representation of mining in general in the novel, is how to elucidate a peculiar endorsement of both modern and premodern ideas about the value and societal impact of economic activity." He further notes that "while

eighteenth-century critiques of luxury are inextricable from the structural economic changes occurring in the period, they often mirror the critiques of antiquity and the Middle Ages in their focus on the individual psychology or even physiology of the consumer." See Matt Erlin, *Necessary Luxuries*, 177 and 179.

17. Michaela Haberkorn contrasts the cave in Novalis's *Ofterdingen,* which constitutes a space of metaphysical insights, with the Platonic cave: "In der Szene wird die Höhle als Ort der tieferen Erkenntnis und Einsicht dargestellt. . . . Es findet geradezu eine Umkehrung von Platos Höhlengleichnis statt, in welchem die Höhle als Ort der unzureichenden Erkenntnis und der Schatten als Gegenpol zur klaren Erkenntnis an der Oberfläche im Tageslicht gezeichnet wird. So wird der Lichtmetaphorik der Aufklärung eine Metaphorik der Dunkelheit zur Kennzeichnung intellektueller Klarheit und Einsicht beigefügt." [In this scene, the cave is depicted as a space of deeper knowledge and insight. . . . This constitutes virtually a reversal of Plato's allegory of the cave, in which the cave is depicted as the space of insufficient insight and of the shadows as the antithesis to the clear insight on the surface in the day light. Thus, to the light imagery of the Enlightenment, imagery of the dark is added to indicate intellectual clarity and insight.] Michaela Haberkorn, *Naturhistoriker und Zeitenseher: Geologie und Poesie um 1800. Der Kreis um Abraham Gottlob Werner (Goethe, A. v. Humboldt, Novalis, Steffens, G. H. Schubert),* Regensburger Beiträge zur deutschen Sprach- und Literaturwissenschaft 87 (Frankfurt am Main: Peter Lang, 2004), 223–24.
18. Mahoney explains that "the ancient German mining traditions held that minerals actually grew underground; this provided a counterbalance to the notion of inanimate nature predominant in Western science and philosophy since Descartes." See Dennis Mahoney, "Human History as Natural History in *The Novices of Sais* and *Heinrich von Ofterdingen,*" *Historical Reflections/Réflexions Historiques* 18, no. 3, The Eighteenth Century and Uses of the Past (Fall 1992): 111–124, 116. Mahoney refers in this statement to Ziolkowski who points out that the "belief that minerals and metals grow and, in the process, become increasingly refined[. . .]provided the basis for alchemy." See Ziolkowski, *German Romanticism and Its Institutions*, 30.
19. A similar fantastic garden appears also in the fairy tale told by the poet Klingsohr, whom Heinrich meets later on in his travels: "Am herrlichsten nahm sich auf dem großen Platze vor dem Pallaste der Garten aus, der aus Metallbäumen und Krystallpflanzen bestand, und mit bunten Edelsteinblüthen und Früchten übersäet war" (*Schriften* I, 291) [What stood out most gloriously was the garden in the large square before the palace, a garden consisting of metal trees and crystal plants, bestrewn with flowers and fruits of sparkling jewels (*Henry*, 121)]. The parallel of the jewel garden that the miner recalls to the one in the fairy tale realm highlights the magical quality of the mineral world, which it has due to its association with alchemistic practices.
20. The merging of the organic and inorganic in the connection between the mineral and the plant world reflects the teachings of Johann Wilhelm Ritter (1776–1810), one of Novalis's teachers at the mining academy in Freiberg. Robert J. Richards notes that Novalis "studied with Johann Wilhelm Ritter . . . , who introduced him to the more outré aspect of natural science, particularly electrical physiology and alchemy." See Robert J. Richards, *The Romantic Conception of Life: Science and Philosophy in the Age of Goethe* (Chicago and London: University of Chicago Press, 2002), 35. For a detailed analysis of Novalis's ideas of procreation in the context of Ritter's work, see Jocelyn Holland, *German Romanticism and Science: The Procreative Poetics of Goethe, Novalis, and Ritter* (New York: Routledge, 2009).
21. As Mahoney states, "Novalis chose to emphasize the bonds of love linking together all living creatures. At the same time he also advocated an envisioned 'moralization of nature,' whereby the world is enhanced and beautified through human work." See Mahoney, "Human History as Natural History," 112.

22. Veronica G. Freeman points out the connection between the self and nature that the blue flower establishes. She states that "[t]his prototypical romantic metaphor turns inward to subjectively focus on the self and its relation to love and nature." See Veronica G. Freeman, *The Poetization of Metaphors in the Work of Novalis*, Studies on Themes and Motives in Literature 78 (Frankfurt am Main: Peter Lang, 2006), 113.
23. Christopher Clason highlights how the Early Romantics dissolved the boundaries of species categories. He explains that "Friedrich Schlegel's transcendental dynamics of combination, which he proposed in his programmatic 116th 'Athenäum-Fragment,' helped to catalyze the dissolution of the proper scientific boundaries—as they had been held for centuries—with regard to the 'Chain of Being' principle. Perhaps the most ardent disciples of this manipulation of species categorization were the early Romantics, whose science anthropomorphized animals, plants, and rocks. Novalis, Tieck, and others were responsible for the creation of a 'New Mythology,' which included a revised understanding of nature and the hierarchy of animate creatures and inanimate objects." See Christopher Clason, "Automatons and Animals: Romantically Manipulating the Chain of Being in E.T.A. Hoffmann's 'Der Sandmann' and *Kater Murr*," in *Romanticism and Beyond: A Festschrift for John F. Fetzer*, ed. Clifford A. Bernd, Ingeborg Henderson, and Winder McConnell, California Studies in German and European Romanticism and in the Age of Goethe 2 (Frankfurt am Main: Peter Lang, 1996), 115–132, 115.
24. Irene Bark points out the permeability of the self and nature in Novalis's works: "Angesprochen wird ein Verhältnis zwischen Mensch und Natur, in welchem eine strikte Dualität zwischen Geistigem und Materiellem, zwischen Subjektivem und Objektivem, zwischen dem Ich und der Natur zugunsten einer Beziehung aufgelöst scheint, deren belebende Impulse zur Herstellung eines wechselseitigen Übergangs zwischen Innen- und Außenwelt führen." [What is addressed is a relationship between human and nature, in which the strict duality between spiritual and material, between subjective and objective, between the self and nature seems dissolved in favor of a relationship, whose vitalizing impulses lead to establishing a reciprocal transition between inner and outer world.] See Irene Bark, *"Steine in Potenzen": Konstruktive Rezeption der Mineralogie bei Novalis*, Hermaea Germanistische Forschungen, Neue Folge 88 (Tübingen: Max Niemeyer, 1999), 250.

ACKNOWLEDGMENTS

The editors would like to express their deep gratitude to Professor Larry Peer of Brigham Young University, Executive Director of the International Conference on Romanticism, our friend and mentor, for his steadfast support for this book and many other endeavors that mean so much to the international academic community of Romanticists. We would also like to thank the organizers of the 2016 International Conference on Romanticism, Professor William Davis and Professor Jared Richman at Colorado College, for their generosity in providing us four sessions during which we were able to plan this project.

Our work has been a labor of love, with its fair share of challenges and trials, and we have found nothing "automatic" in the world of academic research. However, both of us, Michael Demson and Christopher R. Clason, have enjoyed the love and support of our partners in life, Professor Audrey Murfin and Professor Stacey Hahn. We thank them wholeheartedly for their selflessness and generosity during the production of this volume, and it is to them that we wish to dedicate our work, with gratitude and love.

BIBLIOGRAPHY

Adenis, Jules and Octave Gastineau. *La Czarine: Drame en cinq actes et huit tableaux.* Paris: Michel Lévy, 1868.

Agamben, Giorgio. *The Use of Bodies.* Translated by Adam Kotsko. Stanford: Stanford University Press, 2015.

Agrippa, Henry Cornelius. *Three Books of Occult Philosophy or Magic*, Edited by Willis F. Whitehead. Chicago: Hahn & Whitehead, 1898. Reprint, New York: AMS Press.

Ahmed, Sara. *Queer Phenomenology: Orientations, Objects, Others.* Durham and London: Duke University Press, 2006.

Allen, George. "The History of the Automaton Chess-Player in America." In Daniel Willard Fiske, *The Book of the First American Chess Congress*, 420–484. New York: Rudd and Carleton, 1859.

Altick, Richard D. *The Shows of London.* Cambridge and London: Belknap Press of Harvard University Press, 1978.

Anastaski, Elena. "The Trials and Tribulations of the *revenants*: Narrative Technique and the Fragmented Hero in Mary Shelley and Théophile Gautier." *Connotations* 16, no. 1–3 (2006/7): 26–46.

Ancient Origins. "Armillary Spheres: Following Celestial Objects in the Ancient World." Accessed October 28, 2019. http://www.ancient-origins.net/artifacts-ancient-technology/armillary-spheres-following-celestial-objects-ancient-world-004025.

André, Johann and Cuinet-Dorbeil. *Gesänge aus dem Singspiele, Das Automat: in einem Aufzuge.* Hamburg: J. M. Michaelsen, 1787.

Anonymous. "Letter from a father to his daughter at boarding-school." Dublin: Printed for J. Williams, 1774.

Anonymous. *Man Superior to Woman; or, a Vindication of Man's Natural Right of Sovereign Authority over the Woman.* London: T. Cooper, 1739.

Austen, Jane. *Emma.* Edited by Richard Cronin and Dorothy McMillan. Cambridge: Cambridge University Press, 2005.

Austin, Linda M. "Elaborations of the Machine: The Automata Ballets." *Modernism/modernity* 23, no. 1 (2016): 65–87.

Babington, Bruce and Peter Evans. "Matters of Life and Death in Powell and Pressburger's *The Tales of Hoffmann.*" In *A Night in at the Opera: Media Representations of Opera*, edited by Jeremy Tambling. London: John Libbey & Company Ltd., 1994.

Bailey, Brian. *The Luddite Rebellion.* New York: NYU Press, 1998.

Balfour, Ian. *The Rhetoric of Romantic Prophecy.* Stanford: Stanford University Press, 2002.

Ballaster, Ros. "Fakiry: The Oriental Tale." *Great Writers Inspire.* Licensed as Creative Commons BY-NC-SA (2.0 UK). Accessed April 3, 2019. http://writersinspire.org/content/fakiry-oriental-tale.

Bark, Irene. *"Steine in Potenzen": Konstruktive Rezeption der Mineralogie bei Novalis.* Hermaea Germanistische Forschungen, Neue Folge 88. Tübingen: Max Niemeyer, 1999.

Barker, Mary. *A Welsh Story in Three Volumes.* London: Hookham and Carpenter, 1798.

Barnaby, Andrew. "'After the Event': Freud's Uncanny and the Anxiety of Origins." *The Psychoanalytic Quarterly* 84, no. 4 (2015): 975–1006.

Barnard, Frederick M. "Particularity, Universality, and the Hebraic Spirit: Heine and Herder." *Jewish Social Studies* 43, no. 2 (1981): 121–136.

Baudelaire, Charles. "Le morale de joujou." In *Œuvres complètes*, edited by Claude Pichois. Vol. 1. Paris: Gallimard, Bibliothèque de la Pléiade, 1975.

———. *The Painter of Modern Life and Other Essays.* Translated by Jonathan Mayne. London: Phaidon Press, 1964.

———. "The Philosophy of Toys." *Essays on Dolls.* Translated by Idris Parry and Paul Keegan, 15–25. New York: Syrens/Penguin, 1995.

Baudry, Jean-Louis. "The Apparatus: Metaphysical Approaches to the Impression of Reality in Cinema." In *Film Theory and Criticism*, edited by Gerald Mast, Marshall Cohen, and Leo Braudy. New York and Oxford: Oxford University Press, 1992.

Beach, Edward Allen. *The Potencies of God(s): Schelling's Philosophy of Mythology.* Albany: SUNY Press, 1994.

Beck, Heinrich Christian. *Die Schach-Maschine: Ein Lustspiel in 4 Aufzügen.* Vienna: Johann Baptist Wallishaußer, 1798.

Benesch, Klaus. *Romantic Cyborgs: Authorship and Technology in the American Renaissance.* Amherst: University of Massachusetts Press, 2002.

Benhamou, Reed. "From Curiosité to Utilité: The Automaton in Eighteenth-Century France." *Studies in Eighteenth-Century Culture* 17 (1987): 91–105.

Benjamin, Walter. *The Arcades Project.* Translated by Howard Eiland and Kevin McLaughlin. Cambridge and London: Belknap Press of Harvard University Press, 1999.

———. *Berlin Childhood Around 1900.* Translated by Howard Eiland. Cambridge: Harvard University Press, 2006.

———. "E.T.A. Hoffmann und Oskar Panizza." In *Gesammelte Schriften.* Vol. 2, no. 1, 641–648. Frankfurt am Main: Suhrkamp, 1991.

———. "On the Concept of History." In *Walter Benjamin: Selected Writings*, edited by Howard Eiland et al. Translated by Harry Zohn. Vol. 4, 389–400. Cambridge: Harvard University Press, 2006.

———. "Theses on the Philosophy of History." In *Illuminations.* Translated. by Harry Zohn, 255–266. New York: Harcourt, 1955.

———. "Über den Begriff der Geschichte." In *Gesammelte Schriften.* Vol. 1, no. 2, 691–704. Frankfurt am Main: Suhrkamp, 1991.

Bennett, Jane. *Vibrant Matter: A Political Ecology of Things.* Durham: Duke University Press, 2009.

Berkeley, Richard. *Coleridge and the Crisis of Reason.* New York: Palgrave Macmillan, 2007.

Berkley, James. "Post-Human Mimesis and the Debunked Machine: Reading Environmental Appropriation in Poe's 'Maelzel's Chess-Player' and 'The Man That Was Used Up.'" *Comparative Literature Studies* 41, no. 3 (Oct. 2004): 356–376.

Bernasconi, Robert. "Kant and Blumenbach's Polyps: A Neglected Chapter in the History of the Concept of Race." In *The German Invention of Race*, edited by Sara Eigen and Mark Larrimore, 73–90. Albany: SUNY Press, 2006.

Bernstein, J. F., ed. *Classic and Romantic German Aesthetics.* Translated by Stefan Bird-Pollan. Cambridge and New York: Cambridge University Press, 2003.

Bernstein, Susan. "It Walks: The Ambulatory Uncanny." *MLN* 118, no. 5 (2003): 1111–1139.

Blain, Virginia, Patricia Clements, and Isobel Grundy, eds. *The Feminist Companion to Literature in English: Women Writers from the Middle Ages to the Present* (London: Batsford, 1990).

Blumenbach, Johann Friedrich. *Über den Bildungstrieb und das Zeugungsgeschäfte.* Göttingen: Johann Christian Dietereich, 1781.

Böhme, Hartmut. *Natur und Subjekt*. Frankfurt am Main: Suhrkamp, 1988.
Boie, Bernhild. "Der zärtliche Haubenstock und die schöne Automate: Zur weiblichen Kunstfigur der Romantik." *Seminar* 20, no. 4 (1984): 246–261.
Bollacher, Martin. "Johann Gottfried Herder et la conception de l'humanisme." *Les Études philosophiques* 3 (1998): 291–304.
Braine, Sheila E. *The Turkish Automaton: A Tale of the Time of Catherine the Great of Russia*. London: Blackie and Son, 1898.
Bresnick, Adam. "Prosopoetic Compulsion: Reading the Uncanny in Freud and Hoffmann." *The Germanic Review* 71, no. 2 (1996): 114–132.
Brewster, Sir David. *Letters on Natural Magic: Addressed to Sir Walter Scott*. London: John Murray, 1838. 5th ed. 1842.
Brougham, Henry. *An Inquiry into the Colonial Policy of the European Powers*. Edinburgh: D. Willison, 1803.
Bryson, Norman. "Géricault and Masculinity." In *Visual Culture: Images and Interpretations*, edited by Norman Bryson, Michael Ann Holly, and Keith Moxey, 228–260. Hanover: Wesleyan University Press, 1994.
Caillat, Gérald. *The Missing Manuscript: The Story of the Tales of Hoffmann, by Jacques Offenbach*. Documentary. Idéale Audience International, Production Company, and Sept/Arte 2004 (ARTE France). Television.
Calhoon, Kenneth S. *Fatherland: Novalis, Freud, and the Discipline of Romance*. Kritik: German Literary Theory and Cultural Studies. Detroit: Wayne State University Press, 1992.
Canguilhem, Georges. *Knowledge of Life*. Edited by Paola Marrati and Todd Meyers. Translated by Stefanos Geroulanos and Daniela Ginsburg. New York: Fordham University Press, 2008.
Cartwright, Mary Strickland. *A Memoir of the Life, Writings, and Mechanical Inventions of Edmund Cartwright*. London: Saunders and Ottley, 1843.
Castle, Terry. *The Female Thermometer: Eighteenth-Century Culture and the Invention of the Uncanny*. New York and Oxford: Oxford University Press, 1995.
Chadwick, Mary. "Making Space for Wollstonecraft: Mary Barker's *A Welsh Story*." *Romantic Textualities: Literature and Print Culture, 1780–1840* 22 (Spring 2017): 21–35.
Chapuis, Alfred and Edouard Gélis. *Le Monde des automates. Étude historique et technique*. 2 vols. Paris: Slatkine, 1928.
Charlton, Mary. *Rosella, or Modern Occurrences. A Novel in four volumes*. London: Minerva Press, 1799.
Chen, Mel Y. *Animacies*. Durham: Duke University Press, 2012.
Chenique, Bruno. *La folie d'un monde*. Paris: Éditions Hazan, 2006.
Citron, Marcia. *Opera on Screen*. New Haven and London: Yale University Press, 2000.
Cixous, Hélène. "Fiction and its Phantoms: A Reading of Freud's *Das Unheimliche* (The 'Uncanny')." *New Literary History* 7, no. 3 (1976): 525–548 and 619–645.
Clarke, Norma. *Ambitious Heights: Writing, Friendship, Love. The Jewsbury Sisters, Felicia Hemans and Jane Welsh Carlyle*. London: Routledge, 1990.
Clason, Christopher. "Automatons and Animals: Romantically Manipulating the Chain of Being in E.T.A. Hoffmann's 'Der Sandmann' and *Kater Murr*." In *Romanticism and Beyond: A Festschrift for John F. Fetzer*, edited by Clifford A. Bernd, Ingeborg Henderson, and Winder McConnell, 116–132. California Studies in German and European Romanticism in the Age of Goethe 2. Frankfurt am Main: Peter Lang, 1996.
Clément, Charles. *Gericault: Étude Biographique et Critique*. Paris: Didier & Cie, 1879.
Coeckelbergh, Mark. *New Romantic Cyborgs: Romanticism, Information Technology, and the End of the Machine*. Cambridge: MIT Press, 2017.
Colman, George. *The Dramatick Works of George Colman*. 4th edition. London: T. Beckett, 1777.

Coleridge, Samuel Taylor. *The Collected Works of Samuel Taylor Coleridge*. Vol. 11. Edited by H.J. Jackson and J.R. de J. Jackson. Bollingen Series. London and Princeton: Princeton University Press, 1995.

———. *Coleridge: Selected Poems*. Edited by Richard Holmes. New York: HarperCollins, 1996.

Coski, R. Christopher. "Condillac: Language, Thought, and Morality in the Man and Animal Debate." *French Forum* 28, no. 1 (2003): 57–75.

Crawford, Katherine. "Designed Women: Gender and the Problem of Female Automata." *History and Technology* 30 (2014): 261–268.

Curran, Stuart. "The Scientific Grounding of *Frankenstein*." In *Mary vs. Mary*, edited by Lilla Maria Crisafulli, 283–292. Naples: Liguori Editore, 2001.

d'Alembert, Jean-Baptiste le Rond. "Androide," "Automate." In *Encyclopédie, ou dictionnaire raisonné des sciences, des arts et des métiers, etc.*, edited by Denis Diderot et al. Vol. 1. ARTFL Encyclopédie Project, edited by Robert Morrissey et al. (Spring 2016): 448–451, 896–897. Accessed March 28, 2019. https://encyclopedie.uchicago.edu/.

———. "Automaton." In *The Encyclopedia of Diderot & d'Alembert Collaborative Translation Project*. Translated by Nelly S. Hoyt and Thomas Cassirer. Ann Arbor: Michigan Publishing, University of Michigan Library, 2003. Accessed March 28, 2019. http://hdl.handle.net/2027/spo.did2222.0000.140.

Dallmayr, Fred. "Truth and Diversity: Some Lessens from Herder." *Journal of Speculative Philosophy*, N.S. 11, no. 2 (1997): 101–124.

Davis, Lennard. "Nude Venuses, Medusa's Body, and Phantom Limbs." In *The Body and Physical Difference*, edited by David Mitchell and Sharon Snyder, 51–70. Ann Arbor: University of Michigan Press, 1997.

Decremps, Henri. *The Conjurer Unmasked: Or, La Magie Blanche Dévoilée, Being a Clear and Full Explanation of All the Surprizing Performances Exhibited as Well in This Kingdom as on the Continent*. Translated by T. Denton. London: C. Stalker, 1785.

———. *La magie blanche dévoilée ou explication des tours surprenants*. Paris: Langlois, 1784.

DeGabriele, Peter. "Sympathy for the Sovereign: Sovereignty, Sympathy, and the Colonial Relation in Edward Gibbon's 'The Decline and Fall of the Roman Empire.'" *The Eighteenth Century* 53, no. 1 (2012): 1–22.

de Man, Paul. *Aesthetic Ideology*. Minneapolis: University of Minnesota Press, 1984.

Demson, Michael. "Puppets, Automata, and Machinery: Counter-Currents of Transnational Romanticism in Hawthorne's Fiction." In *Nathaniel Hawthorne in the College Classroom*, edited by Christopher G. Diller and Samuel Coale, 247–258. AMS Studies in the Nineteenth Century. Vol. 60. Norwalk, CT: AMS Press, 2016.

Des Chene, Dennis. *Spirits and Clocks: Machine and Organism in Descartes*. Ithaca: Cornell University Press, 2001.

Disraeli, Isaac. *Curiosities of Literature*. 4th ed. London: Murray and Highley, 1798.

Doyon, André and Lucien Liaigre. *Jacques Vaucanson : Mécanicien de Génie*. Paris: Presses Universitaires de France, 1966.

Draude, Claude. "Intermediaries: Reflections on Virtual Humans, Gender, and the Uncanny Valley." *AI and Society* 26 (2011): 319–327.

Dupl, B. A. *The History and Analysis of the Supposed Automaton Chess Player of M. De Kempelen*. Boston: Hillard, Gray and Co., 1826.

Edgeworth, Richard Lovell. *Memoirs of Richard Lovell Edgeworth*. Edited by Maria Edgeworth. London: Richard Bentley, 1844.

Eitner, Lorenz. *Géricault: His Life and Work*. Ithaca: Cornell University Press, 1983.

Engelstein, Stefani. *Anxious Anatomy: The Conception of the Human Form in Literary and Naturalist Discourse*. Albany: SUNY Press, 2008.

———. "Love or Knowledge: Sexual Epistemology in Fichte and Kleist." *Germanic Review* 92, no. 4 (Fall 2017): 368–387.

———. *Sibling Action: The Genealogical Structure of Modernity.* New York: Columbia University Press, 2017.

Erlin, Matt. *Necessary Luxuries: Books, Literature, and the Culture of Consumption in Germany, 1770–1815.* Signale: Modern German Letters, Cultures, and Thought. Ithaca: Cornell University Press, 2014.

Eschenmayer, Carl Adolph. *Versuch die scheinbare Magie des thierischen Magnetismus aus physiologischen und psychischen Gesezen zu erklären.* Stuttgart und Tübingen: J. G. Cotta'schen Buchhandlung, 1816.

Evans, Henry Ridgely. *The Old and the New Magic.* Chicago: Open Court Publishing Company, 1909.

Fabian, Johannes. *Time and the Other: How Anthropology Makes its Object.* New York: Columbia University Press, 2014.

Fanon, Frantz. *Wretched of the Earth.* Translated by Richard Philcox. New York: Grove Press, 1961. Reprint, 2004.

Ferguson, Adam. *An Essay on the History of Civil Society.* Dublin: Boulter Grierson, 1767.

Fetzer, John Francis. "Romantic Irony." In *European Romanticism: Literary Cross-Currents, Modes, and Models*, edited by Gerhart Hoffmeister, 19–36. Detroit: Wayne State University Press, 1990.

Fichte, Johann Gottlieb. *Das System der Sittenlehre, Gesamtausgabe der Bayerischen Akademie der Wissenschaften.* Edited by Reinhard Lauth and Hans Jacob. Stuttgart-Bad Cannstatt: Friedrich Frommann Verlag, 1965.

Foster, William. "Who Wrote Hartly House?" *Bengal Past and Present* 15, pt. 2, no. 30 (1917): 28–29.

Freeman, Kathryn. *British Women Writers and the Asiatic Society of Bengal, 1785–1835: Re-Orienting Anglo-India.* New York: Ashgate, 2014.

Freeman, Veronica G. *The Poetization of Metaphors in the Work of Novalis.* Studies on Themes and Motives 78. Frankfurt am Main: Peter Lang, 2006.

Freud, Sigmund. "Das Unheimliche." *Imago: Zeitschrift für Anwendung der Psychoanalyse auf die Geisteswissenschaften* 5 (1919): 297–324.

———. *Gesammelte Werke: chronologisch geordnet.* 18 vols. Frankfurt am Main: S. Fischer Verlag, 1968–1978.

———. "The Uncanny." In *The Standard Edition of the Complete Psychological Works*, XVII: 217–256. Translated and edited by James Strachey. London: Hogarth Press and Institute of Psycho-Analysis, 1955.

———. The Uncanny." In *The Uncanny,* translated by David McClintock with an introduction by Hugh Haughton, 123–162. New York, Toronto, et al.: Penguin, 2003.

Fried, Michael. *Absorption and Theatricality.* Chicago: University of Chicago Press, 1988.

———. *Courbet's Realism.* Chicago: University of Chicago Press, 1990.

———. "Géricault's Romanticism." In *Géricault*, edited by Régis Michel, 641–659. Paris: Documentation Française, 1996.

———. *Menzel's Realism.* New Haven and London: Yale University Press, 2002.

Gallagher, Catherine. *The Body Economic: Life, Death, and Sensation in Political Economy and the Victorian Novel.* Princeton: Princeton University Press, 2006.

Germano, William. *BFI Film Classics: The Tales of Hoffmann.* London: Palgrave Macmillan, 2013.

Germer, Stefan. "Pleasurable Fear: Géricault and uncanny trends at the opening of the nineteenth century." *Art History* 22, no. 2 (June 1999): 159–183.

Gibbes, Phebe. *Hartly House, Calcutta* (1789). Edited by Michael Franklin. New Delhi: Oxford University Press, 2007.

———. *Zoriada: or, Village Annals.* London: T. Axtell, Royal Exchange, 1786. *ECCO.* http://find.galegroup.com/ecco/retrieve.do?sgHitCountType=None&sort=Author&tabID=T001&prodId=ECCO&resultListType=RESULT_LIST&searchId=R1&searchType=AdvancedSearchForm&currentPosition=17&qrySerId=Locale%28en%2C%2C%29%3AFQE%3D%280X%2CNone%2C7%29zoriada%3AAnd%3ALQE%3D%28BA%2CNone%2C124%292NEF+Or+0LRH+Or+2NEK+Or+0LRL+Or+2NEI+Or+0LRI+Or+2NEJ+Or+0LRK+Or+2NEG+Or+0LRF+Or+2NEH+Or+0LRJ+Or+2NEM+Or+0LRN+Or+2NEL+Or+0LRM%24&retrieveFormat=MULTIPAGE_DOCUMENT&userGroupName=miami_richter&inPS=true&contentSet=ECCOArticles&&docId=CB3330616719&retrieveFormat=MULTIPAGE_DOCUMENT&docLevel=FASCIMILE&workId=CB3330616719&relevancePageBatch=CB130616717&showLOI=&contentSet=&callistoContentSet=ECLL&docPage=article&hilite=y.

Gisborne, Thomas. *An Enquiry into the Duties of the Female Sex*. London: T. Cadell, 1798.

Goethe, Johann Wolfgang von. *Goethes Briefe*. Vol. 2. Hamburger Ausgabe. Hamburg: Christian Wegner, 1962.

Gold, Helmut. *Erkenntnisse unter Tage: Bergbaumotive in der Literatur der Romantik.* Opladen: Westdeutscher Verlag, 1990.

Gold, Joshua Robert. "The Dwarf in the Machine: A Theological Figure and its Sources." *MLN* 121, no. 5 (December 2006): 1220–1236.

Gottlieb, Eric. *Romantic Globalism: British Literature and the Modern World Order, 1750–1830.* Columbus: Ohio State University Press, 2014.

Grant, Iain Hamilton. *Philosophies of Nature after Schelling.* London and New York: Continuum, 2006.

Gregory, Cedric. E. *A Concise History of Mining.* Revised edition. Lisse, Netherlands: A. A. Balkema, 2001.

Green, Katherina Sobba. *The Courtship Novel 1740–1820: A Feminized Genre.* Lexington: University Press of Kentucky, 1998.

Grenby, M. O. *The Child Reader: 1700–1840.* Cambridge: Cambridge University Press, 2011.

Grenville, Bruce, ed. *The Uncanny: Experiments in Cyborg Culture.* Vancouver: Arsenal Pulp Press, 2002.

Grigsby, Darcy Grimaldo. *Extremities: Painting Empire in Post-Revolutionary France.* New Haven: Yale University Press, 2002.

Grimm, Friedrich Melchior. *Correspondance littéraire, philosophique, et critique*. Vol. 13. Paris: Chez Furne, 1830.

Grimstad, Paul. "Antebellum AI: 'Maelzel's Chess-Player' and Poe's Reverse Constraints." *Poetics Today* 31, no. 1 (Spring 2010): 107–125.

Grundy, Isobel. "'the barbarous character we give them': White Women Travelers Report on Other Races." *Studies in English Eighteenth Century Culture* 22 (1992): 73–86.

———. "(Re)discovering Women's Texts." In *Women in Literature in Britain, 1700–1800,* edited by Vivien Jones, 179–96. New York: Cambridge University Press, 2000.

Grusin, Richard. "Introduction." In *The Non-Human Turn*, edited by Richard Grusin, i–xxvi. University of Minnesota Press, 2015.

Gurney, Edmund, Frederic W. H. Myers, and Frank Podmore. *Phantasms of the Living.* 2 vols. London: Rooms of the Society for Psychical Research, 1886.

Gurton-Wachter, Lily. "Toward an Interdisciplinary Sympathy," *Literature Compass*. Accessed October 28, 2019. https://onlinelibrary.wiley.com/doi/abs/10.1111/lic3.12443. Web.

Guyer, Paul. "Knowledge and Pleasure in the Aesthetics of Schelling." In *Interpreting Schelling: Critical Essays*, edited by Lara Ostaric, 71–90. Cambridge: Cambridge University Press, 2014.

Haberkorn, Michaela. *Naturhistoriker und Zeitenseher: Geologie und Poesie um 1800. Der Kreis um Abraham Gottlob Werner (Goethe, A. v. Humboldt, Novalis, Steffens, G. H. Schubert).*

Regensburger Beiträge zur deutschen Sprach- und Literaturwissenschaft 87. Frankfurt am Main: Peter Lang, 2004.

Hadlock, Heather. *Mad Loves: Women and Music in Offenbach's "Les Contes d'Hoffmann."* Princeton: Princeton University Press, 2014.

Hankins, Thomas L. and Robert J. Silverman. *Instruments and the Imagination*. Princeton: Princeton University Press, 1995.

Hanssen, Beatrice. *Walter Benjamin's Other History: Of Stones, Animals, Human Beings, and Angels*. Berkeley and Los Angeles: University California Press, 1993.

Haraway, Donna. "A Cyborg Manifesto: Science, Technology and Socialist-Feminism in the late Twentieth Century." In *Simians, Cyborgs and Women: The Reinvention of Nature*, 149–181. New York: Routledge, 1991.

Hardenberg, Friedrich von (a.k.a. Novalis). *Henry von Ofterdingen*. Translated by Palmer Hilty. Prospect Heights: Waveland Press, 1990.

———. *The Novices of Sais*. Translated by Ralph Manheim. Illustrations by Paul Klee. Brooklyn, New York: Archipelago Books, 2005.

———. *Schriften: Die Werke Friedrich von Hardenbergs*. Edited by Richard Samuel et al. Vol. 3. Stuttgart: Kohlhammer, 1960.

Haslanger, Andrea. "From Man-Machine to Woman-Machine: Automata, Fiction, and Femininity in Dibdin's *Hannah Hewit* and Burney's *Camilla*." *Modern Philology* 111, no. 4 (2014): 788–817.

Hervey, Elizabeth. *Melissa and Marcia; or the Sisters: A Novel. In Two Volumes*. London: W. Lane, 1788.

Heuser-Kessler, Marie-Luise. *Die Produckitvität der Natur. Schellings Naturphilosophie und das neue Paradigma der Selbstorganisation in den Naturwissenschaften*. Berlin: Duncker und Humblot, 1986.

Hillier, Mary. *Automata & Mechanical Toys*. London: Bloomsbury, 1976.

Hindle, Maurice. "'Vital Matters': Mary Shelley's *Frankenstein* and Romantic Science." *Critical Survey* 2, no. 1 (1990): 29–35.

Hoffmann, E.T.A. *E.T.A. Hoffmann Sämtliche Werke*, edited by Gerhard Allroggen, Hartmut Steinecke, Wulf Segebrecht, Ursula Segebrecht. 6 vols. Frankfurt-on-the-Main: Deutscher Klassiker Verlag, 1985–2004.

Vol. 1 = *Frühe Prosa, Briefe, Tagebücher, Libretti, juristische Schrift: Werke 1794–1813*. Edited by Gerhard Allroggen, Friedhelm Auhuber, Hartmut Mangold, Jörg Petzel, and Hartmut Steinecke. 2003.

Vol. 2/1 = *Fantasiestücke in Callot's Manier*. Edited by Hartmut Steinecke with collab. of Gerhard Allroggen and Wulf Segebrecht. 1993.

Vol. 2/2 = *Die Elixiere des Teufels: Werke 1814–1816*. Edited by Hartmut Steinecke with collab. of Gerhard Allroggen. 1988 [2007].

Vol. 3 = *Nachtstücke, Klein Zaches, Prinzessin Brambilla: Werke 1816–1820*. Edited by Hartmut Steinecke with collab. of Gerhard Allroggen. 1985 [2009].

Vol. 4 = *Die Serapionsbrüder*. Edited by Wulf Segebrecht with collab. of Ursula Segebrecht. 2001.

Vol. 5 = *Lebens-Ansichten des Katers Murr: Werke 1820–1821*. Edited by Hartmut Steinecke with collab. of Gerhard Allroggen. 1992.

Vol. 6 = *Späte Prosa, Briefe, Tagebücher und Aufzeichnungen, juristische Schriften: Werke 1814–1822*. Edited by Gerhard Allroggen, Friedhelm Auhuber, Hartmut Mangold, Jörg Petzel, and Hartmut Steinecke. 2004.

———. "Automata." In *The Best Tales of Hoffmann*, edited with an introduction by E.F. Bleiler, 71–103. Translated by Major Alexander Ewing. New York: Dover, 1967.

———. "Beethoven's Instrumental Music: Translated from E.T.A. Hoffmann's 'Kreisleriana' with an Introductory Note." Translated by Arthur Ware Locke. *The Musical Quarterly* 3, no. 1 (1917).

———. "The Sandman." In *The Best Tales of Hoffmann*, edited with an introduction by E. F. Bleiler, 183–214. Translated by Major Alexander Ewing. New York: Dover, 1967.

———. "The Sandman." In *Tales of E.T.A. Hoffmann*, edited and translated by Leonard Kent and Elizabeth Knight, 137–167. Chicago and London: University of Chicago Press, 1972.

Hofkosh, Sonia. "Disfiguring Economies: Mary Shelley's Short Stories." In *The Other Mary Shelley: Beyond Frankenstein*, edited by Audrey A. Fisch, Anne K. Mellor, and Esther H. Schor, 204–219. New York: Oxford University Press, 1993.

Hölderlin, Friedrich. "Oldest Programme for a system of German Idealism." In *Classic and Romantic German Aesthetics*, edited by J. M. Bernstein, 186–187. Translated by Stefan Bird-Pollan. Cambridge and New York: Cambridge University Press, 2003.

Holland, Jocelyn. *German Romanticism and Science: The Procreative Poetics of Goethe, Novalis, and Ritter*. Routledge Studies in Romanticism. New York and Abingdon, UK: Routledge, 2009.

Hunt, Alastair and Matthias Rudolf. "Introduction: The Romantic Rhetoric of Life." *Romanticism and Biopolitics. Romantic Circles* (December 2012). Accessed March 31, 2019. https://romantic-circles.org/praxis/biopolitics/HTML/praxis.2012.hunt-rudolf.html.

Hunter, John. *A Treatise on the Blood, Inflammation, and Gun-shot Wounds*. London: John Richardson, 1794. Eighteenth Century Collections Online. Gale, CW108342200. Accessed March 31, 2019.

Hutcheon, Linda. *A Theory of Adaptation*. New York and London: Routledge, 2013.

Irani, Lilly. "Difference and Dependence among Digital Workers: The Case of Amazon Mechanical Turk." *The South Atlantic Quarterly* 114, no. 1 (Jan. 2015): 225–234.

Ishiguro, Hiroshi. Hiroshi Ishiguro Laboratories (Osaka, 2016). Accessed March 27, 2019. http://www.geminoid.jp/en/index.html.

———. Miraikan National Museum of Emerging Science and Innovation. Accessed March 27, 2019. http://www.miraikan.jst.go.jp/en/exhibition/future/robot/android.html.

Jentsch, Ernst. "On the Psychology of the Uncanny." Translated by Roy Sellars. *Angelaki* 2, no. 1 (1995): 7–16.

Johnson, Barbara. *Persons and Things*. Cambridge: Harvard University Press, 2008.

Johnson, Claudia. *Equivocal Beings*. Chicago: University of Chicago Press, 1995.

Kang, Minsoo. "Building the Sex Machine: The Subversive Potential of the Female Robot." *Intertexts* 9, no. 1 (2005): 5–22.

———. "The Mechanical Daughter of Rene Descartes: The Origin and History of an Intellectual Fable." *Modern Intellectual History* 14, no. 3 (2016): 633–660.

———. *Sublime Dreams of Living Machines*. Cambridge: Harvard University Press, 2011.

Kant, Immanuel. *Critique of Judgment*. Translated by Werner Pluhar. Indianapolis and Cambridge: Hackett Publishing Co., 1987.

Karamzin, Nikolai. *Travels from Moscow through Prussia, Germany, Switzerland, France and England*. London: J. Babcock, 1802.

Kempelen, Wolfgang von, J. G. Mansfeld, and Heinrich Füger. *Wolfgangs von Kempelen K.k. Wirklichen Hofraths Mechanismus der Menschlichen Sprache: Nebst der Beschreibung seiner sprechenden Maschine; Mechanismus der Menschlichen Sprache*. Vienna: J. V. Degen, 1791.

Khalip, Jacques and Forest Pyle. "Introduction: The Present Darkness of Romanticism." In *Constellations of a Contemporary Romanticism*, edited by Jacques Khalip and Forest Pyle. New York: Fordham University Press, 2016.

Kleist, Heinrich von. "On the Marionette Theatre." In *Essays on Dolls*, translated by Idris Parry and Paul Keegan. New York: Syrens/Penguin, 1995.

———, "Über das Marionettentheater." In *Sämtliche Erzählungen*, 555–563. Frankfurt am Main: Deutscher Klassiker Verlag, 2005.

Klenner, Jens. "Geomorphic Poetics: Mountainous Transformations from E.T.A. Hoffmann to Elfriede Jelinek." PhD diss., Princeton University, 2014.

Körner, Hans. *Auf der Suche nach der 'wahren Einheit': Ganzheitsvorstellungen in der französischen Malerei und Kunstliteratur vom mittleren 17. bis zum mittleren 19. Jahrhundert.* Munich: Wilhelm Fink Verlag, 1988.

Kracauer, Siegfried. "Theory of Film." In *Film Theory and Criticism*, edited by Gerald Mast, Marshall Cohen, and Leo Braudy. New York and Oxford: Oxford University Press, 1992.

Küppers, Bernd-Olaf. *Natur als Organismus: Schellings frühe Naturphilosophie und ihre Bedeutung für die moderne Biologie.* Frankfurt: Klostermann, 1992.

Landes, Joan. "The Anatomy of Life: An Eighteenth-Century Perspective." In *Genesis Redux: Essays in the History and Philosophy of Artificial Life*, edited by Jessica Riskin, 96–118. Chicago: University of Chicago Press, 2007.

———. "'Tableaux quasi-vivants' and Simulacra: Anatomy and Design in Vaucanson's Automata." In *L'Automate: modèle métaphore machine merveille*, 189–206. Pessac: Presses Universitaires de Bordeaux, 2013.

Leder, Drew. *The Absent Body.* Chicago: University of Chicago Press, 1990.

Lee, Christina. The Legend of the Christian Arab Madonna in Cervantes' Don Quijote," *Revista Canadiense de Estudios Hispánicos* 32, no. 1 (Autumn 2007): 105–121. Accessed March 31, 2019. https://www.jstor.org/stable/27764178?seq=2#page_scan_tab_contents.

Lenoir, Timothy. *The Strategy of Life: Teleology and Mechanics in Nineteenth-Century German Biology.* Chicago and London: University of Chicago Press, 1982.

Leppert, Richard D. *Music and Image: Domesticity, Ideology and Socio-Cultural Formation in Eighteenth-Century England.* Cambridge: Cambridge University Press, 1988.

Leuven, Adolphe de. *L'automate de Vaucanson.* Paris: Henriot et Cie, 1840.

Levitt, Gerald M. *The Turk, Chess Automaton.* Jefferson, NC: McFarland and Company, 2000.

Liu, Catherine. *Copying Machines: Taking Notes for the Automaton.* Minneapolis: University of Minnesota Press, 2000.

Locke, John. *Some Thoughts Concerning Education.* Oxford: Clarendon, 2000.

Löhr, Robert. *The Chess Machine.* Translated by Anthea Bell. New York: Penguin, 2007.

London, April. *Women and Propriety in the Eighteenth-Century English Novel.* New York: Cambridge University Press, 1999.

Löwy, Michael. *Fire Alarm: Reading Walter Benjamin's "On the Concept of History."* Translated by Chris Turner. London and New York: Verso, 2005.

Luynes, Charles-Philippe d'Albert de. *Mémoires du Duc de Luynes sur la cour de Louis XV (1735–1758).* Edited by L. Dussieux et al. Vol. 2. Paris: Firmin Didot, 1860–1865.

Mahoney, Dennis. *Friedrich von Hardenberg (Novalis).* Sammlung Metzler 336. Stuttgart: Metzler, 2001.

———. "Human History as Natural History in *The Novices of Sais* and *Heinrich von Ofterdingen.*" *Historical Reflections/Réflexions Historiques* 18, no. 3 (Fall 1992): 111–124.

Markley, A. A. "Mary Shelley's 'New Gothic': Character Doubling and Social Critique in the Short Fiction." *Gothic Studies* 3, no. 1 (2001): 15–23.

———. "'The Truth in Masquerade': Cross-Dressing and Disguise in Mary Shelley's Short Stories." In *Mary Shelley's Fictions: From Frankenstein to Falkner*, edited by Michael Eberle-Sinatra and Nora Crook, 109–126. New York: St. Martin's Press, 2000.

Marx, Karl and Friedrich Engels. *Capital.* Translated by Samuel Moore and Edward Aveling. Moscow: Progress Publishers, 2015.

———. *Das Kapital.* In *Werke*, edited by Institut für Marxismus-Leninismus beim ZK der SED. Vol. 23, no. 1. East Berlin: Dietz, 1962.

Massumi, Brian. *Parables for the Virtual: Movement, Affect, Sensation.* Durham: Duke University Press, 2002.

Matthews, Bruce. "Schelling in the Anthropocene: A New Mythology of Nature." *Symposium: Canadian Journal of Continental Philosophy* 19, no. 1 (2015): 94–105.

———. *Schelling's Organic Form of Philosophy: Life as the Schema of Freedom.* Albany: SUNY Press, 2011.

Mbembe, Achille. "Necropolitics." Translated by Libby Meintjes. *Public Culture* 15, no. 1 (2003): 11–40.

McClelland, Charles E. *State, Society, and University in Germany 1700–1914.* Cambridge: Cambridge University Press, 1980.

McGavran, James. "'Insurmountable Barriers to Our Union': Homosocial Male Bonding, Homosexual Panic, and Death on the Ice in Frankenstein," *European Romantic Review* 11, no. 1 (2000): 46–67.

Mellor, Anne K. *Mary Shelley: Her Life, Her Fiction, Her Monsters.* New York: Routledge, 1989.

Messenger, Ann. *His and Hers, Essays in Restoration and Eighteenth-Century Literature.* Lexington: University Press of Kentucky, 1986.

Metz, Christian. "The Imaginary Signifier." In *Film Theory and Criticism*, edited by Gerald Mast, Marshall Cohen, and Leo Braudy. New York and Oxford: Oxford University Press, 1992.

Metzner, Paul. *Crescendo of the Virtuoso: Spectacle, Skill, and Self-Promotion in Paris during the Age of Revolution.* Berkeley: University of California Press, 1998.

Mitchell, David and Sharon Snyder. *Narrative Prosthetics.* Ann Arbor: University of Michigan Press, 2001.

Mitchell, Robert. *Experimental Life: Vitalism in Romantic Science and Literature.* Baltimore: Johns Hopkins University Press, 2013.

———. "Suspended Animation, Slow Time, and the Poetics of Trance." *PMLA* 126, no. 1 (2011): 107–122.

Montag, Warren. "The 'Workshop of Filthy Creation': A Marxist Reading of Frankenstein." In *Frankenstein: Complete Authoritative Text with Biographical and Historical Contexts, Critical History, and Essays from Five Contemporary Critical Perspectives,* edited by Johanna M. Scott. Boston: St. Martin's Press, 1992: 300–311.

Mori, Masahiro. "The Uncanny Valley." Translated by Karl F. MacDorman and Takashi Minato. *Energy* 7, no. 4 (1970): 33–35.

Morse, Hosea Ballou. *The Chronicles of the East India Company Trading to China, 1635–1834.* Vol. 5. Cambridge: Clarendon, 1929.

Mulvey, Laura. "Visual Pleasure and Narrative Cinema." In *Film Theory and Criticism*, edited by Gerald Mast, Marshall Cohen, and Leo Braudy. New York and Oxford: Oxford University Press, 1992.

Müller, Dieter. "Zeit der Automate. Zum Automatenproblem bei Hoffmann." *Mitteillungen der E. T. A. Hoffmann-Gesellschaft* 12 (1966): 1–10.

Muri, Allison. *The Enlightenment Cyborg: A History of Communications and Control in the Human Machine, 1660–1830.* Toronto: University of Toronto Press, 2007.

Nemser, Daniel. *Infrastructures of Race: Concentration and Biopolitics in Colonial Mexico.* Austin: University of Texas Press/Border Hispanisms, 2017.

Nicolai, Friedrich. *Beschreibung einer Reise durch Deutschland und die Schweiz, im Jahre 1781* (Berlin and Stettin: n.p., 1787). Accessed March 28, 2019. https://babel.hathitrust.org/cgi/pt?id=nyp.33433066654074;view=1up;seq=7.

Nochlin, Linda. "Géricault, or the Absence of Women." *October* 68 (Spring 1994): 45–59.

———. *The Body in Pieces: Fragment as Metaphor.* London: Thames & Hudson, 2001.

Nussbaum, Felicity A. *Torrid Zones: Maternity, Sexuality, and Empire in Eighteenth-Century English Narratives.* Baltimore: Johns Hopkins University Press, 1996.

Offenbach, Jacques. *The Tales of Hoffmann.* London Symphony Orchestra conducted by Julius Rudel, 1972. Hamburg: Deutche Grammophon GmbH. Audio CD.

Ostaric, Lara. "The Concept of Life in Early Schelling." In *Interpreting Schelling: Critical Essays*, edited by Lara Ostaric, 48–70. Cambridge: Cambridge University Press, 2014.

Park, Julie. "Pains and Pleasure of the Automaton: Frances Burney's Mechanics of Coming Out." *Eighteenth-Century Studies* 40, no. 1 (2006): 23–49.

Parkin, Chris. "Armillary Sphere." Museum of the History of Science. Accessed October 28, 2019. https://www.youtube.com/watch?v=AaWuJHQL-bQ.

Peters, Diana Stone. "E.T.A. Hoffmann: The Conciliatory Satirist." *Monatshefte* 66, no. 1 (1974): 55–73.

Pinto-Correia, Clara. *The Ovary of Eve: Egg and Sperm and Preformation*. Chicago: University of Chicago Press, 1997.

Pippin, Robert. "Authenticity in Painting." *Critical Inquiry* 31 (Spring 2005): 575–598.

Plato. *Ion*. Translated by Paul Woodruff. In *Complete Works*, edited by John M. Cooper, 937–949. Indianapolis and Cambridge: Hackett Publishing Co., 1997.

Poe, Edgar Allan. "Maelzel's Chess-Player." In *The Complete Works*, edited by James A. Harrison. Vol. 14, 6–37. New York: AMS, 1965.

Pope, Alexander. *The Poetry and Prose of Alexander Pope*. Edited by Aubrey Williams. New York: Houghton Mifflin, 1969.

Porter, James. "Foreword." In *The Body and Physical Difference*, edited by David Mitchell and Sharon Snyder, xiii–xiv. Ann Arbor: University of Michigan Press, 1997.

Pratt, Mary Louise. *Imperial Eyes: Travel Writing and Transculturation*. London and New York: Routledge, 1992.

Preisendanz, Wolfgang. *Humor als dichterische Einbildungskraft: Studien zur Erzählkunst des poetischen Realismus*. Munich: Fink, 1976.

Puttfarken, Thomas. "David's *Brutus* and Theories of Pictorial Unity in France." *Art History* 4, no. 3 (September 1981): 291–304

———, *The Discovery of Pictorial Composition: Theories of Visual Order in Painting 1400–1800*. New Haven: Yale University Press, 2000.

———, *Roger de Piles' Theory of Art*. New Haven: Yale University Press, 1985.

Rabin, Nathan. "I'm sorry for coining the phrase 'Manic Pixie Dream Girl.'" *Salon*, July 16, 2014. Accessed March 24, 2019. http://www.salon.com/2014/07/15/im_sorry_for_coining_the_phrase_manic_pixie_dream_girl/.

Racknitz, Joseph Friedrich Freiherr von. *Ueber den Schachspieler des Herrn von Kempelen, nebst einer Abbildung und Beschreibung seiner Sprachmachine*. Leipzig: J. G. Müller, 1784. 2nd ed. Leipzig: J.G.I. Breitkopf, 1789.

Rajan, Balachandra. "Feminizing the Feminine: Early Women Writers on India." In *Romanticism, Race and Imperial Culture, 1780–1834*, edited by Sonia Hofkosh and Alan Richardson, 49–72. Bloomington: Indiana University Press, 1996.

Rajan, Tilottama. "Dis-figuring Reproduction: Natural History, Community, and the 1790s Novel." *CR: The New Centennial Review* 2, no. 3 (Fall 2002): 211–252.

———. "Evolution and its Resistances: Transferences between Disciplines in Schelling's and Hegel's Systems," *Symposium: Canadian Journal of Continental Philosophy* 19, no. 1 (2015): 154–175.

———. "Organicism." *English Studies in Canada* 30, no. 4 (December 2004): 46–50.

———. "A Peculiar Community: Mary Shelley, Godwin, and the Abyss of Emotion." In *Romanticism and Emotion*, edited by Richard Sha and Joel Faflak, 145–170. Cambridge: Cambridge University Press, 2014.

Reill, Peter H. *Vitalizing Nature in the Enlightenment*. Berkeley and Los Angeles: University of California Press, 2005.

Reilly, Kara. *Automata and Mimesis on the Stage of Theatre History*. New York: Palgrave MacMillan, 2011.

Reininger, Alice. *Wolfgang von Kempelen: A Biography*. Translated by Peter Waugh. New York: Columbia University Press, 2012.

Richards, Robert J. *The Romantic Conception of Life: Science and Philosophy in the Age of Goethe*. Science and its Conceptual Foundations. Chicago and London: University of Chicago Press, 2002.

Richter, Jean Paul Friedrich. "Auswahl aus des Teufels Papieren." In *Jean Pauls Werke*. Vol. 4, 1–301. Berlin: Gustav Hempel, 1870.

Rigby, Kate. "'Mines Aren't Really Like That': German Romantic Undergrounds Revisited." In *German Ecocriticism in the Anthropocene*, edited by Caroline Schaumann and Heather I. Sullivan. Literature, Cultures, and the Environment, 111–128. New York: Palgrave Macmillan, 2017.

Risinger, Jacob. "Coleridge, Politics, and the Theory of Life." *Studies in English Literature* 55, no. 3 (Summer 2015): 647–667.

Riskin, Jessica. "The Defecating Duck, or, The Ambiguous Origins of Artificial Life." *Critical Inquiry* 29, no. 4 (Summer 2003): 599–633.

———. "Eighteenth-Century Wetware." *Representations* 83, no. 1 (Summer 2003): 97–125.

———. *The Restless Clock: A History of the Centuries-Long Argument over What Makes Living Things Tick*. Chicago and London: University of Chicago Press, 2016.

———. *Science in the Age of Sensibility: The Sentimental Empiricists of the French Enlightenment*. Chicago: Chicago University Press, 2002.

Roe, Shirley. *Matter, Life, and Generation: Eighteenth-Century Embryology and the Haller-Wolff Debate*. Cambridge: Cambridge University Press, 1981.

Rosenberg, Jordana. *Critical Enthusiasm: Capital Accumulation and the Transformation of Religious Passion*. Oxford: Oxford University Press, 2011.

Rosenthal, Adam. "The Gift of Memory in Baudelaire's 'Morale du joujou.'" *Nineteenth-Century French Studies* 43, no. 3–4 (2015): 129–143.

Ross, Marlon. "Quest and Conquest." In *Romanticism and Feminism*, edited by Anne Mellor. Bloomington: Indiana University Press, 1988.

Rousseau, Jean-Jacques. *Émile, or, On Education*. Translated by Allan Bloom. New York: Basic Books, 1979.

Ruston, Sharon. *Shelley's Vitality*. New York: Palgrave Macmillan, 2005.

Ryle, Gilbert. *The Concept of Mind*. Chicago: University of Chicago Press, 1949.

Sauer, Lieselotte. "Romantic Automata." In *European Romanticism: Literary Cross-currents, Modes, and Models*, edited by Gerhart Hoffmeister, 287–306. Detroit: Wayne State University Press, 1990.

Schaffer, Simon. "Enlightened Automata." In *The Sciences in Enlightened Europe*, edited by William Clark et al., 126–165. Chicago: University of Chicago Press, 1999.

Schelling, Friedrich Wilhelm Joseph. *Ausgewählte Schriften*. Edited by Manfred Frank. Frankfurt am Main: Suhrkamp Verlag, 1985.

———. "Concerning the Relation of the Plastic Arts to Nature." Translated by Michael Bullock. In Herbert Read, *The True Voice of Feeling: Studies in English Romantic Poetry*, 321–364. London: Faber and Faber, 1953.

———. *Philosophical Investigations into the Essence of Human Freedom*. Translated by Jeff Love and Johannes Schmidt. Albany: SUNY Press, 2006.

———. *Philosophische Untersuchungen über das Wesen der menschlichen Freiheit und die damit zusammenhängenden Gegenstände*. Edited by Thomas Buchheim. Hamburg: Felix Meinter Verlag, 1997.

———. *System of Transcendental Idealism* (1800). Translated by Peter Heath. Charlottesville: University Press of Virginia, 1978.

———. "Treatise on the Relationship of the Real and the Ideal in Nature." Translated by Dale Snow. *International Philosophical Quarterly* 55, no. 22 (2015): 235–250.

———. *Von der Weltseele: eine Hypothese der höheren Physik zur Erklärung des allgemeinen Organismus*. Edited by Jörg Jantzen; *Historisch-Kritische Ausgabe*. Edited by Hans Michael Baumgartner, Wilhelm Jacobs, Jörg Jantzen, and Hermann Krings. Stuttgart: Frommann-Holzboog, 2000.

Schillace, Brandy Lain. "Reproducing Custom: Mechanical Habits and Female Machines in Augustan Women's Education." *Feminist Formations* 25, no. 1 (2013): 111–137.

Schubert, Gotthilf Heinrich. *Ansichten von der Nachtseite der Naturwissenschaft*. Dresden: in der Arnoldischen Buchhandlung, 1808.

Sedgwick, Eve Kosofsky & Adam Frank. "Shame in the Cybernetic Fold: Reading Sylvan Tomkins." *Critical Inquiry* 21, no. 2 (1995): 496–522.

Seymour, Miranda. *Mary Shelley*. New York: Grove Press, 2000.

Sha, Richard. "The Motion behind Romantic Emotion." In *Romanticism and the Emotions*, edited by Joel Faflak and Richard Sha, 19–47. Cambridge: Cambridge University Press, 2014.

Shaftesbury, Anthony Ashley Cooper, Earl of. *A Notion of the Historical Draught or Tabulature of the Judgment of Hercules, According to Prodicus, Lib. II. Xen. de Mem. Soc.* London: A. Baldwin, 1713. Accessed March 31, 2019. http://find.galegroup.com/ecco/infomark.do?contentSet=ECCOArticles&docType=ECCOArticles&bookId=0346000600&type=getFullCitation&tabID=T001&prodId=ECCO&docLevel=TEXT_GRAPHICS&version=1.0&source=library.

Sharpe, Jenny. *Allegories of Empire: The Figure of Woman in the Colonial Text*. Minneapolis: University of Minnesota Press, 1993.

Shaw, Devin Zane. *Freedom and Nature in Schelling's Philosophy of Art*. London and New York: Continuum, 2010.

Shelley, Mary Wollstonecraft. *Frankenstein: Or, The Modern Prometheus*. Edited by Susan Wolfson. London: Longman, 2003.

———. *Frankenstein: Annotated for Scientists, Engineers, and Creators of All Kinds*. Edited by David Guston, Ed Finn, and Jason Scott Robert. Cambridge: MIT Press, 2018.

———. *The Journals of Mary Shelley*. Edited by Paula R. Feldman and Diana Scott-Kilvert. New York: Oxford University Press, 1987.

———. *The Letters of Mary Wollstonecraft Shelley*. Edited by Betty T. Bennett. Baltimore: Johns Hopkins University Press, 1980.

———. *Mary Shelley: Collected Tales and Stories*. Edited by Charles E. Robinson. Baltimore: Johns Hopkins University Press, 1976.

Shelley, Percy Bysshe. *The Collected Poems of Percy Bysshe Shelley*. Edited by Donald H. Reiman and Neil Fraistat. New York: W. W. Norton & Co., 2002.

———. "A Philosophical View of Reform." In *Shelley: The Major Works*, 636–674. Oxford: Oxford University Press, 2009.

Sieferle, Rolf Peter. *Der unterirdische Wald: Energiekrise und Industrielle Revolution*. Die Sozialverträglichkeit von Energiesystemen. Edited by Klaus Michael Meyer-Abich, Bertram Schefold, and Carl Freidrich von Weizsäcker. Munich: C. H. Beck, 1982.

Sitter, Zak. "Inorganic Intentions: Organizing Form from *Frankenstein* to de Man." *Literature Compass* (October 13, 2016): 655–662.

Smith, Allan Lloyd. "'This Thing of Darkness': Racial Discourse in Mary Shelley's Frankenstein," *Gothic Studies* 6, no. 2 (2004): 208–222.

Smith, Amy Elizabeth. "Roles for Readers in Mary Wollstonecraft's *A Vindication of the Rights of Woman*." *SEL* 32, no. 3 (1992): 555–570.

Snow, Dale. *Schelling and the End of Idealism*. Albany: SUNY Press, 1996.

Southey, Robert. *The Curse of Kehama*, London: Longman, 1810.

Standage, Tom. *The Turk: The Life and Times of the Famous Eighteenth-Century Chess-Playing Machine*. New York: Walker & Co., 2002.

Stedman, John Gabriel. *Narrative of a Five Years' Expedition; against the Revolted Negroes of Surinam in Guiana.* London: J. Johnson and J. Edwards, 1796.

Steigerwald, Joan. "Epistemologies of Rupture: The Problem of Nature in Schelling's Philosophy." *Studies in Romanticism* 41, no. 4 (2002): 545–584.

Stephen, James. *The Crisis of the Sugar Colonies: or, an Enquiry into the Objects and Probable Effects of the French Expedition to the West Indies.* London: J. Hatchard, 1802.

Stern, Lesley. "An Instance of Operality." In *Between Opera and Cinema*, edited by Jeongwon Joe and Rose Theresa, 39–57. New York and London: Routledge, 2002.

Sterne, Jonathan. *The Audible Past: Cultural Origins of Sound Reproduction.* Durham: Duke University Press, 2003.

Sullivan, Heather I. "Dirty Nature: Ecocriticism and Tales of Extraction—Mining and Solar Power—in Goethe, Hoffmann, Verne, and Eschbach." *Colloquia Germanica* 44, no. 2 (2011): 111–131.

———. "Ruins and the Construction of Time: Geological and Literary Perspectives in the Age of Goethe." *Studies in Eighteenth-Century Culture* 30 (2001): 1–30.

Tartar, Maria M. "E.T.A. Hoffmann's 'Der Sandmann': Reflection and Romantic Irony." *MLN*, 95, no. 3 (1980): 585–608.

Teltscher, Kate. *India Inscribed: European and British Writing on India 1600–1800.* Delhi: Oxford University Press, 1995.

Thicknesse, Philip. *The Speaking Figure and the Automaton Chess Player, Exposed and Detected.* London: Printed for John Stockdale, 1784.

Todd, Janet. *Sensibility: An Introduction*. New York: Methuen, 1986.

Toepfer, Georg. "Organismus." In *Historisches Wörterbuch der Biology: Geschichte und Theorie der biologischen Grundbegriffe.* Stuttgart and Weimar: J. B. Metzler, 2011.

Tresch, John. *The Romantic Machine: Utopian Science and Technology after Napoleon*. Chicago: University of Chicago Press, 2012.

———. "The Science and Politics of the Fantastic Automaton." *French Historical Studies* 34, no. 1 (2011): 87–123.

Trillini, Regula Hohl. *The Gaze of the Listener: English Representations of Domestic Music-Making.* Amsterdam: Rodopi, 2008.

Trop, Gabriel. "The Aesthetics of Schelling's Naturphilsophie." *Symposium: Canadian Journal of Continental Philosophy* 19, no. 1 (2015): 140–152.

Turner, Cheryl. *Living by the Pen: Women Writers in the Eighteenth Century*. London: Routledge, 1992.

Uerlings, Herbert. "Novalis in Freiberg: Die Romantisierung des Bergbaus—Mit einem Blick auf Tiecks *Runenberg* und E.T.A. Hoffmanns *Bergwerke zu Falun*." *Aurora* 56 (1996): 57–77.

Ure, Andrew. *The Cotton Manufacture of Great Britain: Systematically Investigated, and Illustrated by 150 Original Figures, Engraved on Wood and Steel; with an Introductory View of Its Comparative State in Foreign Countries, Drawn Chiefly from Personal Survey.* London: H. G. Bohn, 1836.

Vaucanson, Jacques de. *An Account of the Mechanism of an Automaton, or Image Playing on the German-Flute.* Translated by John T. Desgauliers. London: T. Parker, 1742.

———. *Le mécanisme du fluteur automate: presenté à Messieurs de l'Académie royale des sciences.* Paris: J. Guérin, 1738.

Vlasopolos, Anca. "Mary Wollstonecraft's Mask of Reason in *A Vindication of the Rights of Woman*." *Dalhousie Review* 60, no. 3 (1980): 462–471.

Voltaire. "Discours en vers sur l'homme." In *Oeuvres complètes de Voltaire*. Vol. 9. Paris: Garnier frères (1877), 379–428.

Voskuhl, Adelheid. *Androids in the Enlightenment: Mechanics, Artisans, and Cultures of the Self.* Chicago and London: University of Chicago Press, 2013.

Walker, Gavin. "Primitive Accumulation and the Formation of Difference: On Marx and Schmitt." *Rethinking Marxism* 23, no. 3 (2011): 384–404.

Warminski, Andrzej. "Introduction: Allegories of Reference." In *Aesthetic Ideology*, edited by Andrzej Warminski. Minnesota: University of Minnesota Press, 1996.

Weatherby, Leif. *Transplanting the Metaphysical Organ: German Romanticism between Leibniz and Marx.* New York: Fordham University Press, 2016.

Weheliye, Alexander G. *Habeas Viscous: Racializing Assemblages, Biopolitics, and Black Feminist Theories of the Human*. Durham: Duke University Press, 2014.

Wetmore, Alex. *Men of Feeling in Eighteenth-Century Literature: Touching Fiction.* New York: Palgrave Macmillan, 2013.

———. "Sympathy Machines: Men of Feeling and the Automaton." *Eighteenth-Century Studies* 43, no. 1 (2009): 37–54.

Wheatstone, Charles. *The Scientific Papers of Sir Charles Wheatstone.* London: Taylor and Francis, 1879.

Wiegleb, Johann Christian. *Die natürliche Magie.* Berlin and Stettin: Friedrich Nicolai, 1789.

Wilcox, Kirstin. "Vindicating Paradoxes: Mary Wollstonecraft's 'Woman.'" *Studies in Romanticism* 48, no. 3 (2009): 447–467.

Williams, Helen Marie. *Letters Written in France*. Edited by Neil Fraistat. Peterborough: Broadview Press, 2001.

Willis, Robert. *An Attempt to Analyze the Automaton Chess Player of Mr. de Kempelen*. Edinburgh: A. Constable, 1821.

Windisch, Karl Gottlieb von. *Briefe über den Schachspieler des Herrn von Kempelen.* Preßburg: A. Löwe, 1783.

Wirth, Jason. *Schelling's Practice of the Wild: Time, Art, Imagination.* Albany: SUNY Press, 2015.

Wolf, Rebecca. *Friedrich Kaufmanns Trompeterautomat: Ein musikalisches Experiment um 1810.* Stuttgart: Franz Steiner, 2011.

Wolfe, Cary. *What Is Posthumanism.* Minneapolis: University of Minnesota Press, 2010.

Wollstonecraft, Mary. *Letters Written in Sweden, Norway, and Denmark*. Edited by Tone Brekke and John Mee. Oxford: Oxford University Press, 2009.

———. *The Vindications: The Rights of Men and The Rights of Woman*. Edited by D. L. Macdonald and Kathleen Scherf. Peterborough, ON: Broadview, 1997.

———. *A Vindication of the Rights of Woman.* Edited by Deirdre Shauna Lynch. New York: Norton, 2009.

Wood, Gabby. *Living Dolls: A Magical History of the Quest for Mechanical Life.* London: Faber & Faber, 2003.

Youngquist, Paul. "Lyrical Bodies: Wordsworth's physiological aesthetics." *European Romantic Review* 10, no. 1–4 (1999): 152–162. Accessed March 30, 2019. https://doi.org/10.1080/10509589908570073.

Zammito, John. "The Lenoir Thesis Revisited: Blumenbach and Kant." *Studies in the History and Philosophy of Biological and Biomedical Sciences* 43 (2012): 120–132.

Ziolkowski, Theodore. *German Romanticism and Its Institutions*. Princeton: Princeton University Press, 1990.

NOTES ON CONTRIBUTORS

FREDERICK BURWICK (University of California, Los Angeles), research professor, has enjoyed visiting positions in Germany at the universities of Würzburg, Siegen, Göttingen, and Bamberg. He has been named Distinguished Scholar by both the British Academy (1992) and the Keats-Shelley Association (1998). With an interdisciplinary approach to literature, Burwick developed his courses at UCLA to explore the interactions of literature with art, science, music, and theater. Author and editor of thirty-four books, one hundred sixty articles, and numerous reviews, his research is dedicated to problems of perception, illusion, and delusion in literary representation and theatrical performance. Among his books devoted to performance are *Romantic Drama: Acting and Reacting* (2009), *Playing to the Crowd: London Popular Theatre, 1780–1830* (2011), *British Drama of the Industrial Revolution* (2015), and *British Pirates in Print and Performance* (2015) with Manushag Powell.

CHRISTOPHER R. CLASON (Oakland University), professor emeritus of German, works on German Romanticism and the Middle Ages. He has authored articles on E.T.A. Hoffmann's novels and tales, Romantic narrative and the Gothic novel, and numerous topics in medieval German literature, including *Tristan*, *Parzival*, and animals and nature in literature. He is co-editor of *Literary and Poetic Representations of Work and Labor in Europe and Asia during the Romantic Era* and *Romantic Rapports*, and editor of *E.T.A. Hoffmann: Transgressive Romanticism*. Clason is past President of the International Conference on Romanticism and the International Tristan Society.

MICHAEL DEMSON (Sam Houston State University), associate professor of English, has published articles in *European Romantic Review*, *Romanticism*, *Romantic Circles Praxis*, among others, and co-edited *Commemorating Peterloo: Violence, Resilience, and Claim-making in the Romantic Era* (2019). He is currently completing a monograph, *The Blight of Corn*, on agrarian radicalism and Romanticism, a project supported by an American Council of Learned Societies Fellowship. His non-fiction graphic novel, *Masks of Anarchy: From Percy Shelley to the Triangle Shirtwaist Factory Fire*, was published by Verso Books in 2013.

STEFANI ENGELSTEIN (Duke University), professor of German and chair of the German Department, is the author of *Sibling Action: The Genealogical Structure of Modernity* (2017) and *Anxious Anatomy: The Conception of the Human Form in Literary and Naturalist Discourse* (2008), and the co-editor of *Contemplating Violence: Critical Studies in Modern German Culture* (2011). Her articles on German and British literature, science, and philosophy have appeared in such journals as *Critical Inquiry*, the *PMLA*, the *German Quarterly*, and *Philosophy Today*. Engelstein has been a visiting scholar at the Max Planck Institute for the History of Science and at the Center for Literary and Cultural Research, both in Berlin. She is currently working on two book projects, *The Entanglements of the Organism* and *The Opposite Sex: A History*.

PETER ERICKSON (Colorado State University), assistant professor of German, researches the intersection of literature, art, and religion in the eighteenth and nineteenth centuries. For his dissertation at the University of Chicago, he used archival research to show how a debate surrounding the role of religious conversion in the public sphere influenced the development of the German *Bildungsroman*. His publications include articles on the work of Friedrich Schiller, Jacob Friedrich Abel, E.T.A. Hoffmann, Gotthold Ephraim Lessing, and the theologian Gotthilf Samuel Steinbart.

KATHRYN FREEMAN (University of Miami), professor of English, is the author of *Blake's Nostos: Fragmentation and Nondualism in The Four Zoas* (1997) and *A Guide to the Cosmology of William Blake* (2017). Her *Women Writers and the Asiatic Society of Bengal, 1785–1835: Re-Orienting Anglo-India* (2014) investigates the literary relationship between women writers of the late eighteenth century and the Orientalist transmission of Vedic texts.

ERIN M. GOSS (Clemson University) is associate professor of English. Her scholarship focuses on eighteenth- and nineteenth-century writing by and about women, especially as it intersects with contemporary gender politics. She is the author of *Revealing Bodies: Anatomy, Allegory, and the Grounds of Knowledge in the Long Eighteenth Century* (2013) and has recently edited a collection of essays on *Jane Austen and Comedy* (2019). Her current monograph is titled "Echo's Trap: British Women, Complicity, and Acts of Conscience, 1759–1835."

LENORA HANSON (New York University), assistant professor of English, has published in *Essays in Romanticism*, *theory@buffalo* and *Multitudes* and has an essay on Giacomo Leopardi in *European Romantic Review*. Her current project, *Riotous Life: The Rhetoric and Politics of Romantic Organisms,* investigates poetic analogies forged between nonhuman life, reproduction, and the politics of disruption throughout the eighteenth and nineteenth century.

WENDY C. NIELSEN (Montclair State University), associate professor of English, has authored *Women Warriors in Romantic Drama* (2012) and scholarly essays on Romantic-era automata, theater, the drama of the French Revolution, Jean-Jacques Rousseau, Olympe de Gouges, Johann Wolfgang von Goethe, Elizabeth Inchbald, Charlotte Corday, and Boadicea. Her current book project, *Motherless Creations: Fictions of Artificial Life, 1650–1900*, examines the ways in which British, German, French, and American literature intervenes in the scientific discourse of mothering, parenting, and procreation in order to argue for the ways in which writing mediates fictions about the future of the posthuman.

ASHLEY SHAMS (University of St. Thomas), associate professor of French, teaches undergraduate courses in language, poetry, and civilization. Her research interests include the study of liminal figures in gothic fiction and literature of the fantastic as well as the intersection of Romanticism, dance, and opera. She most recently published a co-authored article on Gautier's Romantic ballet *Giselle* and his novella *La morte amoreuse* in *Romantic Rapports*, edited by Larry H. Peer and Christopher R. Clason. She served as president of the International Conference on Romanticism, 2014–2016.

KATE SINGER (Mount Holyoke College), associate professor of English, has written articles on Percy Shelley, Maria Jane Jewsbury, Mary Robinson, Letitia Landon, and Charlotte Smith that have appeared in *Studies in Romanticism*, *European Romantic Review*, *Literature Compass*, *Romantic Circles Praxis*, and *Essays in Romanticism*. The editor of the *Pedagogies* section of the *Romantic Circles* website, including the online journal *Pedagogy Commons*, she has published on digital pedagogy in *JiTP* with another essay forthcoming on Letitia Landon in *Pedagogy*. She edited, along with Nanora Sweet, a special issue of *Women's Writing* on Felicia Hemans and is currently finishing a book entitled *Romantic Vacancy: Affect, Gender, and Radical Speculation*.

CHRISTINA M. WEILER (University of North Carolina, Chapel Hill), teaching assistant professor of German, has lived and taught in Germany, Spain, Ireland, and the United States. Her research interests lie in metaphor, cognition, environmental studies, and memory studies with a focus on German literature of the long eighteenth century. Her publications include articles on Johann Gottfried Herder and E.T.A. Hoffmann. Her current book project is titled *The Romantic Roots of Cognitive Poetics: A Comparative Study of Poetic Metaphor in Herder, Novalis, Wordsworth, Coleridge, and Shelley*.

INDEX

Note: Figures are indexed in italic.